The Fermat Diary

Fine Hall, November 1995

The Fermat Diary

C. J. Mozzochi

2000 *Mathematics Subject Classification.* Primary 11D41.

The cover photo is the house where Fermat was born. The front was altered in the 18th century and is not identical to the façade in Fermat's time. The photograph is courtesy of Prof. Dr. Klaus Barner.

Excerpts from "Fame by Numbers", by Ian Katz, April 8, 1995, are provided by *The Guardian*, copyright © 1995 by *The Guardian*. Reprinted with permission.

Excerpts from "How a Gap in the Fermat Proof was Found", by Gina Kolata, January 31, 1995, are provided by the *New York Times*, copyright © 1995 by the *New York Times*. Reprinted by permission.

Excerpts from the script of the BBC's *Horizon* series broadcast presentation of *Fermat's Last Theorem*, January 15, 1996, are provided by the BBC, copyright © 1996 by BBC Worldwide Publishers. Reprinted with permission.

Excerpts from the Introduction to "Modular Elliptic Curves and Fermat's Last Theorem," *Ann. of Math* **142** (1995), pp. 443–551, are provided by Andrew Wiles, copyright © 1995 by Andrew Wiles. Reprinted with permission.

Library of Congress Cataloging-in-Publication Data
Mozzochi, C. J.
 The Fermat diary / C. J. Mozzochi.
  p. cm.
 Includes bibliographical references and index.
 ISBN 0-8218-2670-0
 1. Fermat's Last Theorem. I. Title.
QA244.M69 2000
512′.74—dc21   00-030629

Copying and reprinting. Individual readers of this publication, and nonprofit libraries acting for them, are permitted to make fair use of the material, such as to copy a chapter for use in teaching or research. Permission is granted to quote brief passages from this publication in reviews, provided the customary acknowledgment of the source is given.

Republication, systematic copying, or multiple reproduction of any material in this publication is permitted only under license from the American Mathematical Society. Requests for such permission should be addressed to the Assistant to the Publisher, American Mathematical Society, P. O. Box 6248, Providence, Rhode Island 02940-6248. Requests can also be made by e-mail to reprint-permission@ams.org.

© 2000 by the American Mathematical Society. All rights reserved.
The American Mathematical Society retains all rights
except those granted to the United States Government.
Printed in the United States of America.

∞ The paper used in this book is acid-free and falls within the guidelines
established to ensure permanence and durability.
Visit the AMS home page at URL: http://www.ams.org/
10 9 8 7 6 5 4 3 2 1   05 04 03 02 01 00

Dedicated to the memory of my father and mother

Contents

Preface	ix
Chapter 1. February 10, 1994	1
Chapter 2. September 19, 1994	53
Chapter 3. August 9, 1995	97
Cast of Characters	125
List of Photographs	133
Acknowledgements and Notes	137
Appendix A. An Excerpt From the Introduction to Wiles's Annals of Mathematics Paper	151
Appendix B. List of Speakers, Cambridge University, June 1993	163

Appendix C. List of Speakers, Boston University, August 1995	165
Appendix D. Ram Murty's Review	171
Bibliography	179
Index	189

Preface

Fermat's Last Theorem states that if n is an integer greater than 2, the equation $x^n + y^n = z^n$ has no solutions if x, y, and z are restricted to being positive integers.

The French mathematician and jurist, Pierre de Fermat (1601–1665), asserted in 1637 that he possessed a truly marvelous proof, which was too long to write in the margin of his copy of Book II of the *Arithmetica*, written (c. AD 250) by Diophantus of Alexandria. For some 350 years after Fermat's claim no valid proof had been constructed albeit it had attracted the attention of some of the world's greatest mathematicians. In fact a substantial amount of modern mathematics evolved from attempts by experts to construct a valid proof.

It became one of the greatest unsolved problems in all of mathematics.

Perhaps the strong appeal of the problem is the simplicity and elegance of its statement contrasted with the apparent hopelessness of finding an elementary way to establish it.

The first real breakthrough that ultimately led to a correct proof quietly surfaced in the autumn of 1984 at the Mathematisches Forschungsinstitut Oberwolfach near Freiburg, Germany when the German mathematician, Gerhard Frey, stated informally and plausibly that the Shimura-Taniyama conjecture, restricted to semistable elliptic curves, implies Fermat's Last Theorem.[1]

In August 1986, the American mathematician, Kenneth Ribet, following a difficult and speculative plan of attack suggested by the eminent French mathematician, Jean-Pierre Serre, established rigorously the statement made by Frey.

On June 23, 1993 the British-American mathematician, Andrew Wiles, announced in Cambridge, England that he had established the Shimura-Taniyama conjecture for semistable elliptic curves, but on December 4, 1993 he announced that there was a gap in his proof.

On September 19, 1994 Wiles (with the collaboration of his former British student, Richard Taylor) dramatically closed the gap in his proof.

From February 10, 1994 through August 18, 1995, Wiles gave twenty-six, one-hour U.S.A. lectures concerning his proof. Twenty were given at various locations in Princeton, New Jersey, and one each was given at Harvard University, Hunter College, Yale University, New York University, Columbia University, and Boston University. He also gave a lecture at Rutgers University on February 23, 1996.

From August 9, 1995 to August 18, 1995, the Instructional Conference on Number Theory and Arithmetic Geometry was held at Boston University, and thirty-three technical lectures were given there by other experts on various aspects of Wiles's proof.

I attended and tape-recorded each of those lectures with one exception, the lecture given by Wiles in Princeton on February 23, 1994. I took technical notes at those lectures and maintained a journal. Additionally, I possess the negatives for more than three thousand photographs which I took before, during, and after most of those lectures.

Preface   xi

    This book is based on the above-mentioned material, together with my recollections and impressions of events, supplemented by those of others who so kindly shared their recollections and impressions with me.

    This was a most dramatic and exciting time in mathematics, and I have endeavored to capture and record it accurately. Posterity will be my judge.

C. J. Mozzochi, Ph.D.
Princeton, N.J.
December 31, 1999

Chapter 1

February 10, 1994

Summer days in Princeton can be difficult with the searing heat, the stifling humidity and the sudden severe storms, and June 24, 1993 was no exception. I was at work in my summer term office at the Institute for Advanced Study when a secretary informed me that she heard on the radio while driving to work that Andrew Wiles had proved an important result.[1]

I immediately went to the kiosk at Palmer Square and read the headlines in the upper left–hand corner of the front page of the *New York Times*, "At Last, Shout of 'Eureka!' In Age-Old Math Mystery." The two column article, which continued on page D22 where it occupied one-third of the page, announced something startling: Wiles claimed at a conference in Cambridge, England to have proved Fermat's Last Theorem! I was captivated by the significance of his claim, if it proved to be valid after his paper had been properly refereed by a panel of experts, and I just stood there reading the article carefully several times until there was simply nothing more that I could wring out of it.

At tea that afternoon there was a very high level of excitement, but unfortunately many of the faculty members were away at various conferences so there were few experts with whom to discuss the situation. The consensus was that, since Wiles had such an impressive track record, his claim was to be taken very seriously indeed. I overheard one visiting mathematician, who himself had done some impressive research, say that this was in a sense a very depressing turn of events for most research mathematicians, because if Wiles's proof was ultimately found to be correct, no matter what they did during the rest of their careers, it would never measure up to a proof of Fermat's Last Theorem, whereas until his announcement, there were many first-rate mathematicians "in the same boat" with Wiles.

Within hours of Wiles's original announcement, Karl Rubin of Ohio State and Ken Ribet of Berkeley, who were present at the Cambridge Conference, sent separate e-mails to people in their respective institutions, and these messages were eventually distributed throughout the mathematics community, giving brief sketches of his proof (Ribet's was the longer and contained the most detail). For the time being, those e-mails were all anyone could read to get any insight into what Wiles had done by way of a proof.[2]

In fact, in the case of Ribet this happened by mistake. The message that Ribet sent was addressed to the "local" people on the Berkeley number theory e-mail list. He intended his message to go to the graduate students and faculty in number theory at Berkeley. What happened, however, was that the local list had an address line that resulted in a large number of MSRI people getting the message, and this increased the probability that the message would get distributed widely.[3] It did!

Approximately two weeks after Wiles returned from his lectures in Cambridge, a reception was held for him in the Fine Hall Common Room where, instead of tea, champagne was served in his honor. The president of Princeton University, Harold Shapiro, was in attendance.

On July 12th, Enrico Bombieri, who had attended the conference where Wiles made his claim, informed me during tea at the Institute that although he had not yet seen a manuscript, the proof that Wiles had outlined in his three lectures appeared to him to present a solid

1. February 10, 1994

attack on the problem. Later, after the gap in the proof had been found, Bombieri would say when approached on the subject, "I am rooting for Wiles."[4]

On July 28, 1993 at 8:00 PM in the Palace of Fine Arts in San Francisco, a conference, which was sponsored by the Mathematical Sciences Research Institute located in Berkeley, entitled, *Fermat's Last Theorem. The Theorem and Its Proof: An Exploration of Issues and Ideas,* was held to give the ubiquitous "intelligent layperson" some insight into what Wiles and others had done and the significance of their achievements. The 1000 seat auditorium was filled to capacity. The moderator was Will Hearst, a relative of William Randolph Hearst of "Citizen Kane" fame and since 1984 the publisher and editor of the *San Francisco Examiner.* Informal (for the most part elementary but nevertheless very informative) expository talks were presented by Robert Osserman, Lenore Blum, Karl Rubin, Ken Ribet, and John Conway. These talks were followed by a brief panel discussion moderated by Will Hearst. The panel consisted of Lenore Blum, John Conway, Lee Dembart and Ken Ribet. Two musical interludes were provided by Morris Bobrow, who sang mathematical songs written by Tom Lehrer. A video tape recording of the entire conference, lasting approximately one hour and forty minutes, was subsequently made available to the public. The recording contains a brief interview with Andrew Wiles and is supplemented with a booklet edited by Robert Osserman.

In the July/August 1993 issue of the *Notices of the American Mathematical Society,* Ribet provided an elegant but concise two page outline of the proof. On August 3, 1993 Ribet gave a comprehensive lecture on Wiles's proof at George Washington University lasting almost two hours, and on August 15, 1993 at Vancouver, BC, Barry Mazur gave an interesting, forty-five minute lecture on the proof. The video tape recordings from those lectures were very helpful to me and to many others in learning the basic structure of the proof (but not the tortuous details). There was virtually nothing else available to the serious mathematician until the paper by Cox appeared in January of 1994, the paper by Gouvêa appeared in March of 1994, the paper by Rubin and Silverberg appeared in July of 1994 and Wiles ultimately released a full manuscript on October 25, 1994.[5]

Public discussion of the proof, however, grew rapidly. The prestigious Fields Institute for Mathematical Research, located in Toronto, Ontario, Canada, had scheduled for the 1993–94 academic year a conference and a special year devoted to the study of L-functions. After Wiles made his June 1993 claim in Cambridge, the focus of the conference changed and evolved into a Seminar on Fermat's Last Theorem. And during the fall of 1993, participants of the special year studied and lectured on background material needed to understand the proof. A proceedings volume, edited by Kumar Murty, was published based on the talks and discussions of the seminar (see [**81**]).

In 1993, from December 18th through December 21st, the work of Wiles also motivated a conference on elliptic curves and modular forms at the Chinese University of Hong Kong. One of the editors of the resulting volume (see [**15**]), Shing-Tung Yau, told me that because of limited facilities, only approximately sixty participants were allowed to attend. The other editor of the volume, John Coates, had been Wiles's Ph.D. advisor, and they wrote several important papers together (cf. Photograph 7).

All of this activity was the culmination of Wiles's interest in the problem from an early age. The scenario was essentially this.[6] At the age of ten (c. 1963), Wiles found in his local library on Milton Road in Oxford a book that stimulated his interest in Fermat's Last Theorem. The book was entitled *The Last Problem* and had been written by E.T. Bell in 1961. From this time on, and throughout his teenage years, he made numerous elementary attacks on the problem, assuming—with the arrogance that often manifests itself at that age in very highly intelligent, young boys—that Fermat could not have known very much more about it than he did.[7] During college and graduate school, and later as he pursued his career as a professional

1. February 10, 1994

mathematician, he wisely refrained from spending serious research time on the problem, until a colleague informed him in 1986 of the work of Frey, Serre and Ribet that linked the problem to a difficult area of mainstream mathematics, in which he was expert.

In the late 1970's Gerhard Frey (cf. Photograph 43) was discussing with mathematicians who would listen to him his, at the time preliminary, ideas that ultimately evolved into his linking Fermat with Shimura-Taniyama. It was not, however, until the autumn of 1984, when he informally discussed his matured ideas with a small group of mathematicians at the Mathematisches Forschungsinstitut Oberwolfach near Freiburg, Germany deep in the majestic Black Forest, that his ideas began to attract the attention that they deserved of very competent mathematicians, who had the intelligence and the requisite mathematical tools to properly address and exploit his most significant and fertile mathematical insight.[8]

Ribet explained it to me this way:

> I first heard about his ideas when Frey was in Berkeley in the summer of 1981. However, it never sunk into me that he was hoping to prove so simple an implication as "Shimura-Taniyama implies Fermat." Meanwhile, his ideas depended on the truth of a level-lowering principle of the type that I established later, and at the time I thought that such a principle might be false. So I was pretty oblivious to Frey's contribution back then. As far as I can recall, it was first explained to me that Frey was trying to prove "Shimura-Taniyama implies Fermat" in January 1985. I was in Paris then, and some friends had come back from a meeting in Oberwolfach where Frey had lectured and had distributed a very short manuscript.[9]

At this time Frey is somewhat out of the spotlight. Perhaps posterity will more properly recognize him for the substantial contribution that he made toward the solution of the problem.

There is not much about Frey to be found in the literature. Singh and Aczel provide some information in their books; in his book van der Poorten succintly discusses Frey:

> Frey first mentioned his notions at an Oberwolfach meeting, and subsequently reported in detail in a local journal ['Links between stable elliptic curves and certain diophantine equations,' *Ann. Univ. Sarav. Math. Ser.* **1** (1986), pp 1–40]; a later accessible reference is Gerhard Frey, 'Links between solutions of A-B=C and elliptic curves,' in H. P. Schlickewei and E. Wirsing eds., *Number theory, Ulm* 1987 (15th Journées Arithmétiques, Ulm 1987), Springer Lecture Notes **1372** (New York: Springer-Verlag, 1989). But Frey's remarks were not the first to suggest a connection between elliptic curves and Fermat's Last Theorem. At Besançon in 1972, Yves Hellegouarch [cf. Photograph 42] wrote a thesis 'Courbes elliptiques et équation de Fermat.' However, Hellegouarch applies known information about the FLT to elliptic curves rather than using the elliptic curves to learn about Fermat's equation; see 'Points d'ordre $2p^h$ sur les courbes elliptiques' [*Acta Arith.* **26** (1975), pp. 252–263].

Frey has told me that I am helping to perpetuate a legend with the above remark. Frey had studied his curves earlier independently of, but in exactly the context of, the work of Hellegouarch. He points out that his observations at Oberwolfach were not just a random aside, but a natural development of

1. February 10, 1994

his own work. Indeed, the earlier ideas had arisen in the context of studying Ogg's conjecture (later to become Mazur's theorem) on torsion points on rational elliptic curves. At that time the connection to Fermat's Last Theorem was a bad thing, an obstruction. After all, one knew the FLT to be inaccessible, so it was never good to find it intruding into matters one hoped to prove. On the other hand, after Mazur's work it became reasonable to ask whether knowledge of elliptic curves might yield insight into the FLT.[10]

Frey explained it to me this way:

> Thank you for your e-mail. I shall try to answer your questions.
>
> As you indicate it really was not just one moment during which the ideas to relate Fermat with Taniyama appeared, rather the story goes back, as far as I am concerned, to the early seventies. At this time many people in number theory, in particular Hellegouarch, Demjanenko and myself, tried to exclude the existence of torsion points on elliptic curves over **Q**. The methods available at this time led to local considerations and one could not avoid diophantine questions related to Fermat's problem when one tried to globalize these conditions.
>
> I think that Hellegouarch did this first. His results were published rather late ['Points d'ordre $2p^h$ sur les courbes elliptiques', *Acta Arith.* **26** (1975), pp. 252–263] since he was stuck at some point. So I did mostly the same computations independently (as well as Demjanenko). I gave some seminar talks

about this, 1972/73, e.g., at the University of Erlangen. The outcome of these considerations is published in an article in *Arkiv für Mat.*; 1977. I generalized my computations considerably during the publication process, and so it took a long time until publication. This gave me the opportunity to cite Hellegouarch in the paper since I learned about his results while proofreading my paper.

I considered this work as being a little disappointing. To try to solve a question and to come to Fermat's problem is not encouraging. The situation completely changed (and here the heart of the story began to beat) after the publication of the seminal paper of B. Mazur [cf. Photographs 4 and 45] about the Eisenstein ideal in 1976 (see [**70**]). A consequence of his work is that there are no torsion points (up to small exceptions) on elliptic curves over **Q** and (together with results of Serre) that the image of the Galois representations attached to torsion points is as large as possible. Moreover his method to prove his results made it crystal clear how all important the theory of modular curves and forms are. Just as a side remark: One cannot overestimate Mazur's importance in the whole story.

Anyway Mazur's result gave a link not between torsion points and Fermat over the same field but with nontrivial Galois module structures, which is much richer. At this point I became excited. (As far as I know Hellegouarch did not follow here.)

Let me explain the situation from the point of view of an algebraic number theorist: Because of Mazur's result the existence of a solution of FLT would imply that a large Galois extension of **Q** would

1. February 10, 1994

exist which would be nearly unramified over cyclotomic fields. This reminds one of Kummer's approach. But he was handling abelian extensions and so he "could use" Kummer theory and class field theory (which, of course, did not exist at his time).

In our situation the extensions were not solvable, but they were attached to two-dimensional Galois representations. So Langlands philosophy comes to mind; a mod p version was (a little bit vaguely) formulated by Serre about 1972.

I discussed with Ribet these considerations during the late 70's. But since the formulation of Serre's conjecture was not explicit enough, we dropped the subject soon. Nevertheless I continued to study the relations between modular theory and Fermat's conjecture. A related but slightly different angle was published in the paper: Rationale Punkte auf Fermatkurven und getwistete Modulkurven, *Crelle*, 1982. I gave a talk on this subject at Harvard in 1981 and at Berkeley.

Now it became clear to me that the modularity of elliptic curves would be the only possibility to get results on FLT from this line of investigation. On several occasions I mentioned this relation referring to the early formulation of Serre's conjecture. One of these occasions was in the fall of 1984 at Oberwolfach. Here I definitely claimed that Taniyama would imply FLT, and I gave some arithmetical reasons, too. This was not a formal lecture but an informal discussion as is typical for the style and the atmosphere at Oberwolfach.

I was at this time well-aware of the fact that the later-so-called Frey curves were semistable. The necessary results can be found in my paper, 'Some remarks concerning points of finite order on elliptic curves over global fields', *Ark. für Math.* **15** (1977) No.1, or even explicitly stated and proved in the paper 'Rationale Punkte auf Fermatkurven und getwisteten Modulkurven', *J. Reine Angew. Math.* **331** (1982), 185–191 in section 2, pp. 187–88, especially Folgerung on p. 188 and Lemma 1. The result is stated over arbitrary number fields and for general Fermat curve solutions, but it reduces over **Q** and for the special Fermat equation to the desired result outside of the prime 2. For this prime (not essential for Wiles's proof of FLT) one needs indeed a special normalization to get semistability.

Some friends asked me to write up these ideas. A short letter (definitely not a paper) was written. It contained the conjecture: Taniyama implies Fermat, and it gave a short and, in general, wrong sketch of how to prove it. Indeed it gave a proof which works in the easy case called Mazur's result, usually.

I do not have a copy of this letter. I sent it to only a few people, but it got out of control somehow (which was fortunate for mathematics). Very soon it became clear that the sketch of a "proof" of the implication was not sufficient, but remember that this was not for public distribution. Nevertheless in early 1985 it was formulated clearly what was missing (Serre's epsilon-conjecture), and I wrote a paper in *Annales Saraviensis* in which I gave a report on the ideas and on the relations between several conjectures. During a meeting in Oberwolfach in the winter of 1985 I gave such a report, too. At the

1. February 10, 1994

> same time the very stimulating conjecture about elliptic curves of Szpiro [cf. Photograph 17] emerged, and this relates Fermat type equations or more generally the ABC conjecture of Masser-Oesterle [which immediately implies Fermat's Last Theorem] to natural conjectures about arithmetic surfaces via elliptic curves. I am convinced that this line of research will give exciting results.[11]

Frey has been awarded the Gauss Medal of the Scientific Society of Braunschweig. He holds a membership in the Academy of Sciences Göttingen and an honorary Ph.D. from the University of Kassel.

Above we have Frey acknowledging Mazur's impact with regard to his work. Mazur's work on deformations of Galois representations is crucial to Wiles's proof (further acknowledgment by Wiles of other work of Mazur that was important to his proof will be found later in this chapter). Ribet utilizes the results of Mazur on bounds for the torsion of elliptic curves over $\mathbf{Q}$.

Ribet recounts a further contribution by Mazur:

> In August of 1986 I saw Barry Mazur on campus and I said, "Let's go for a cup of coffee" and we sat down for cappuccinos at this café [where the interview was being conducted] and I looked at Barry and I said, "You know, I am trying to generalize what I have done so that we can prove the full strength of Serre's epsilon conjecture," and Barry looked at me and said, "Well you have done it already, all you have to do is add on some extra gamma zero of m structure and run through your argument and it still works, and that gives everything you need." This had never occurred to me as simple as it sounds. I looked at Barry, I looked to my cappuccino, I looked

back at Barry and said, "My God, you're absolutely right!"[12]

On August 13, 1985 Jean-Pierre Serre (cf. Photographs 8 and 9) outlined in a letter to Jean-François Mestre his ideas for justifying Frey's technically-defective argument. Serre indicated that if his conjectures C_1 and C_2 could be proved, then Frey's insight would be rigorously established.[13] These conjectures concern modular forms (mod p) on $\Gamma_0(N)$ and Galois representations.

In August of 1986 Ribet (cf. Photograph 44) established Serre's conjectures, and Fermat was now incorporated into mainstream mathematics. However, it took months before he had the complete proof written down carefully with all of the "bugs" expunged from it, and then, at the request of Serre, he further revised his manuscript during the following academic year (1987–1988) while he was preparing it for publication in the very prestigious mathematics research journal, *Inventiones Mathematicae*, the journal to which Wiles originally submitted his paper before the gap was found in his proof.[14]

The contribution that Ribet made toward the resolution of Fermat's Last Theorem was substantial, for prior to his establishing C_1 and C_2 several well-known experts openly expressed the opinion that they were not true. Further, there is an unexpected, purely technical result (the level-lowering principle) within Ribet's proof that Wiles incorporated strategically within his proof. Finally, we must assume that a mathematician of Serre's stature and reputation would ponder his conjectures very carefully and very thoroughly before releasing them to the mathematics community to be established, so that what Serre left for Ribet to do required mathematical skill and knowledge of the highest level.

In 1989 Ribet was awarded the Prix Fermat jointly with Abbas Bahri.[15] He has been elected to the American Academy of Arts and Sciences as well as to the National Academy of Sciences, and he has

1. February 10, 1994

received an honorary doctorate from Brown University where he received his undergraduate degree.

The very day Wiles heard of Ribet's result he decided to work on establishing Shimura-Taniyama restricted to semistable elliptic curves, but it would take him five years to find the key insight (see Appendix A).

Wiles explained it this way:

> I was at a friend's house sipping tea early in the evening, and he just mentioned casually in the middle of a conversation, "By the way, did you hear that Ken has proved the epsilon conjecture?" And I was just electrified. I knew at that moment the course of my life was changing because this meant that to prove Fermat's Last Theorem I just had to prove the Taniyama-Shimura conjecture. From that moment that was what I was working on. I just knew I would go home and work on the Taniyama-Shimura conjecture.
>
> So it was now known that Taniyama-Shimura implied Fermat's Last Theorem. Now you might ask why can you not do this with elliptic curves and modular forms. Why could you not count elliptic curves, count modular forms, and show they are the same number? Well the answer is people tried, and they never found a way of counting, and this was why this is the key breakthrough, namely, that I found a way to count not the original problem, but the modified problem. I found a way to count modular forms and Galois representations.[16]

How nice it must be to work at that level of mathematics where to complete one's research project one *just* has to establish a conjecture of the depth of Shimura-Taniyama!

He worked in total isolation on the problem, for a substantial part of the time in his attic office in his Princeton home (cf. Photograph 41). Then in January 1993 he took Princeton colleague and friend, Nicholas Katz, into his confidence (cf. Photograph 3).

Wiles was apprehensive about the area of his proof that employed Euler products and the Kolyvagin-Flach method for bounding the order of the Selmer group attached to the symmetric square representation of a modular form. He and Katz decided that the best way to properly and carefully check this portion of the proof would be for Wiles to lecture on it during his 1993 spring term graduate course on algebraic number theory concerning "calculations on elliptic curves," which Katz would attend. This was to be done in such a way that the graduate students would not be aware of the connection of Wiles's lectures with Fermat. Eventually everyone attending the course except Katz drifted away. At the end of the course Wiles felt confident that this portion of his proof was correct.

Previously, in the fall of 1992, Wiles had devoted his graduate course to the theory of deformations of Galois representations, in particular to the work of Mazur, who made a significant contribution to this theory, which plays a crucial role in Wiles's proof. Shou-Wu Zhang (cf. Photograph 16) attended the course, and after it ended he approached Wiles to discuss with him his perceived relationship of this theory to the Fermat problem.[17] Whereupon Wiles suggested that he read Flach and Greenberg (see [**35**], [**43**]). So, late in 1992 there were hints to perceptive people like Zhang of what Wiles was up to.

Also, some of Wiles's applications for financial grants for research from the National Science Foundation gave hints of what he was up to, but few people bother to read such documents even though under

1. February 10, 1994

the Freedom of Information Act they are available to anybody who requests copies of them.

Wiles explained it this way:

> Now I tried to use Iwasawa theory in this context, but I ran into trouble. I seemed to be up against a wall. I just didn't seem to be able to get past it. Well, sometimes when I cannot see what to do next I often come here by the lake [where the interview is being conducted]. Walking has a very good effect in that you're in this state of concentration, but at the same time you are relaxing, you are allowing the subconscious to work for you.
>
> So at the end of the summer of 1991 I was at a conference where John Coates told me about a wonderful new paper of Matthias Flach, a student of his, in which he had tackled a class number formula, in fact exactly the class number formula I needed. Flach, using ideas of Kolyvagin, had made a very significant first step in actually producing the class number formula. At that point I thought this is just what I need, this is tailor-made for the problem. I put aside completely the old approach I had been trying, and I devoted myself day and night to extending his result.
>
> Well, I explained at the beginning of the course that Flach had written this beautiful paper, and I wanted to try to extend it to prove the full class number formula. The only thing I did not explain was that proving the class number formula was most of the way to Fermat's Last Theorem.[18]

He had one more part of his proof to complete. After he completed that portion of the proof, he felt confident to announce his results at the conference at Cambridge University in England in June. At this time he took his colleague, friend and neighbor, Peter Sarnak, into his confidence (cf. Photograph 6).

Wiles explained it this way:

> But there was still a problem. Late in the spring of 1993 I was in this very awkward position. I thought I had got most of the curves to be modular so that was nearly enough to be content to have Fermat's Last Theorem, but there was a problem; there were these few families of elliptic curves that had escaped the net, and I was sitting here at my desk [in his home attic office] in May of 1993 still wondering about this problem, and I was casually glancing at a paper of Barry Mazur's. There was one sentence which made a reference to actually what's a 19th-century construction, and I just instantly realized that there was a trick that I could use, namely, that I could switch from the families of elliptic curves I had been using (I had been studying them using the prime three). I could switch and study them using the prime five. It looked more complicated, but I could switch from these awkward curves that I could not prove were modular to a different set of curves which I had already proved were modular and use that information to just go that one last step. I just kept working out the details and time went by and I forgot to go down to lunch and it got to about teatime and I went down and Nada was very surprised that I had arrived so late. I told her that I believed I had solved Fermat's Last Theorem.

1. February 10, 1994

> I was convinced that I had Fermat in my hands, and there was a conference in Cambridge organized by my advisor, John Coates. I thought that would be a wonderful place (it's my old home town, I'd been a graduate student there) to talk about it if I could get it in good shape.[19]

In June 1993 at the Isaac Newton Institute for Mathematical Sciences at Cambridge University, Wiles gave lectures on Monday the 21st, Tuesday the 22nd and Wednesday the 23rd. The titles of his lectures were *Modular forms, elliptic curves and Galois representations* I., II., III., and on each day the lectures were delivered between 10:00 AM and 11:00 AM.[20]

Li Guo, who had been a Ph.D. student of Ralph Greenberg and whose work is referred to in Wiles's *Annals of Mathematics* paper, was present at the Cambridge Conference when Wiles gave his lectures (see [**45**], [**130**]).

He related to me his observations:

> Ralph Greenberg told me before the first talk that Wiles's lectures would be very important.
>
> During the first lecture the room was filled to two-thirds capacity. At the end of the lecture Wiles said that what he had just talked about that day might have application to the Taniyama-Shimura conjecture.
>
> The second lecture was very technical and very difficult.
>
> During the last lecture the room was filled to capacity with approximately seventy to eighty people present. Half way through the lecture several people took flash photographs of Wiles.

At the end of the talk Wiles wrote the Taniyama-Shimura conjecture as a theorem on the board. And then he wrote:

Corollary $U^p + V^p + W^p = 0$, with U, V, W in $\mathbf{Q} \to UVW = 0 \quad p > 1$.

Several people again took flash photographs of Wiles; in fact I took one myself (cf. Photograph 1). Then there was sustained applause for Wiles.

Ralph Greenberg talked later that day. Before the talk started, Ken Ribet said to him, "Ralph we do not want any more surprises." Whereupon Greenberg replied, " I will be surprised if I can finish my talk." Indeed, Greenberg implied that he might still be too shocked to finish.

After the talks were over for that day, there was a reception in the Fellows Garden of Emmanuel College at Cambridge University. It was a beautiful day, and the weather was perfect.

Champagne was served to the seventy to eighty people who were present, and John Coates led a toast to Wiles. During the course of the reception some people approached Wiles and congratulated him. One person said that now that the psychological barrier has been removed from the problem, there probably will be a simple proof of Fermat in a couple of years.[21]

That night Wiles had dinner at Tickell Arms, an historic pub approximately five miles south of Cambridge, with Ken Ribet, Richard Taylor, Karl Rubin, Fred Diamond, Ehud de Shalit, Barry Mazur, and Enrico Bombieri, who was keen on ordering a fancy bottle of red wine to commemorate the occasion (cf. Photographs 4,14,31,44,45,48,51).

1. February 10, 1994

But the pub did not do much business in red wine so he had to settle for something which was only a local maximum.

Wiles left the conference before noon the next day to visit his family in Oxford. Barry Mazur contacted Gina Kolata, a science writer at the *New York Times*, and Ken Ribet acted as Wiles's press secretary.

Soon after the conference Wiles gave his paper to Barry Mazur, who was an editor of *Inventiones Mathematicae*. Mazur divided the paper into six sections and assigned one section to each of six referees.

Section six (some 60 pages in length) was assigned to Nick Katz, who started to review it with Luc Illusie while both were in Paris for the summer. The section assigned to Katz encompassed essentially the same area of the proof that he and Wiles had very carefully reviewed together the previous spring. Late in August Katz, in the course of reviewing his assigned portion of the paper, raised a technical question to Wiles, which ultimately led Wiles to conclude that there was a gap in the proof. On December 4, 1993 Wiles sent the following e-mail to the mathematics community:

> Subject: Fermat Status
> Date: 4 Dec 93 01:36:50 GMT
>
> In view of the speculation on the status of my work on the Taniyama-Shimura conjecture and Fermat's Last Theorem I will give a brief account of the situation. During the review process a number of problems emerged, most of which have been resolved, but one in particular I have not settled. The key reduction of (most cases of) the Taniyama-Shimura conjecture to the calculation of the Selmer group is correct. However the final calculation of a precise upper bound for the Selmer group in the semistable case (of the symmetric square representation associated to a modular form) is not yet complete as it stands. I

believe that I will be able to finish this in the near future using the ideas explained in my Cambridge lectures.

The fact that a lot of work remains to be done on the manuscript makes it still unsuitable for release as a preprint. In my course in Princeton beginning in February I will give a full account of this work.

Andrew Wiles

A serious controversy developed because, except for the six referees and except for the very few members of Wiles's close inner circle, nobody else had access to Wiles's manuscript, and this fact upset many people. The crux of the criticism was that he announced his claim of Fermat without first having his proof reviewed carefully and that he was behaving improperly by refusing to release his manuscript. However, after it was generally acknowledged that the gap had been fixed, there were nothing but accolades directed toward Wiles and his monumental achievement.

At the time of the controversy Peter Sarnak pointed out that many critics were ignoring the fact that if Wiles had done nothing more than show that there exists an infinity of semistable modular elliptic curves over $\mathbf{Q}$ each associated with a distinct j-invariant, the part of his proof that was rigorous from its very inception, and if he had never done or ever claimed anything else toward Fermat, his work would have immediately been recognized as a very fundamental contribution to mathematics. (Previously Weil showed that there is an infinity of modular elliptic curves over $\mathbf{Q}$, but his curves cannot be shown to be in one-to-one correspondence with distinct j-invariants.)[22]

It should be mentioned that at the time it was not generally known that Wiles had spent an entire semester working very closely

1. February 10, 1994

with Katz carefully checking the deepest and most difficult part of the proof before he announced his claim of Fermat. Also, perhaps posterity will more easily recognize the wisdom of Wiles's decision to not release any portion of the manuscript until he either closed the gap in the proof or until he decided that he had nothing more of substance to say concerning the closing of the gap.

Throughout the controversy Wiles maintained his silence and held his head high. Never once did I observe him to not be in good spirits and in good humor. In fact early in the spring term of 1994 Wiles shared with a small group of us who had joined him for tea the following incident. One day shortly after he had made his Cambridge announcement Wiles was walking along the shore of Lake Carnegie in Princeton, where Einstein used to frequently sail his small boat, when he encountered a layperson walking in the opposite direction. As they passed each other the man said to Wiles, "Oh, I see you have solved a very important problem!" A few months later Wiles again encountered the same man, and as they passed each other the man said to Wiles, "Oh, I see you are having difficulties!"

After Wiles made his claim of Fermat, articles appeared in newspapers such as *The New York Times* and in popular science magazines such as *Scientific American*, and many of those articles were posted on the bulletin boards in the halls of Fine Hall. Not one single article was removed after Wiles had announced that there was a gap in his proof. In fact some of them remained there for several years.

The only public comments Wiles made concerning his proof were those in his Cambridge lectures in June 1993 and those in his e-mail of December 4, 1993, but he did announce that his graduate course for the spring term of 1994 would include lectures by him on various aspects of his proof, with the first lecture to be held in Taplin Auditorium in Fine Hall on February 10, 1994.

Taplin Auditorium was filled to capacity on that cold, dark and dreary winter afternoon. Mathematicians from many institutions were there, and although several reporters tried to crash his lecture, all were denied access essentially on the grounds that he was conducting a class and not a public lecture. At least one television network sent a satellite truck to Princeton, but it soon departed. I was the only person allowed to take photographs, but I was not permitted to use a flash attachment so I had to shoot in extremely poor, available light with a newly released, very high speed (3200 ASA) Kodak professional film.

The auditorium was charged with the excitement and anticipation of the moment. For several months rumors had been circulating, which seemed to exemplify every possible logical outcome. Wiles arrived a few minutes after the announced starting time of the lecture, and he sat with Zeev Rudnick, Nick Katz and Gerd Faltings in a front row. After he was briefly introduced by Yakov Sinai, Wiles walked slowly to the board, and he equally slowly retrieved a piece of chalk. He then turned to the audience, and nonchalantly said in a quiet voice, "Well, in case any of you were wondering, I will say that I still haven't completely resolved the problem that arose in the review process." "... in case any of you were wondering..." that must have been the understatement of the century, and the audience exploded with laughter.

He then proceeded to carefully and thoroughly explain in full technical detail the difficulty concerning the proof of the conjectured upper bound of the cardinality of the Selmer group attached to the symmetric square representation of a modular form that was forcing him to rethink that portion of his proof (cf. Photograph 2).

At the end of his lecture he said the format for his class that term would be that he would lecture each Wednesday at 2:00 PM in Taplin Auditorium, and graduate students in the class would lecture in room 214 each Monday at 2:30 PM on background material needed to establish the proof. After a few lectures, Wiles's lectures were relocated to room 110, which was significantly smaller than Taplin Auditorium.

1. February 10, 1994

A few seconds after Wiles concluded his lecture, out of the corner of my eye I noticed Robert Langlands, Enrico Bombieri, Pierre Deligne, Nick Katz, Gerd Faltings and Peter Sarnak surrounding Wiles at the board (cf. Photographs 3,6,10,11,12,24). It would have made a most striking and a most important historical photograph, and I was out of film. Such is the fate of the unthinking paparazzo!

That semester Wiles gave ten lectures, Brian Conrad two, Chris Skinner three, Nike Vatsal two, and Conrad and Vatsal gave two joint lectures. All were Ph.D. students of Wiles. On April 25th Jerry Tunnell of Rutgers University gave a very carefully prepared lecture on his paper, which plays a fundamental role in Wiles's proof.

When Langlands wrote his important, Steele-Prize-winning monograph on base change, he included some examples. When Tunnell wrote his paper, he was providing yet another important example, employing the results in Langlands's monograph, the joint results of Jacquet, Piatetski-Shapiro and Shalika combined with his own original ideas (cf. Photographs 20–24).[23]

Throughout the term approximately thirty-five to forty people religiously attended Wiles's lectures. Among the faithful were Pierre Deligne, Nick Katz, Gerd Faltings, Peter Sarnak, Jerry Tunnell, Richard Taylor, Shou-Wu Zhang, Larry Washington, Henri Darmon, Fred Diamond, Chris Skinner, Brian Conrad and Nike Vatsal (cf. Photographs 3,6,10–16,18–20,31). Although many people asked questions at these lectures, most of the time Wiles was relentlessly cross-examined by Deligne, Katz and Faltings, and their perceptive questions were most stimulating and illuminating to the other participants. Wiles held his own nicely with no apparent effort. Particularly memorable was the time Wiles calmly and painfully patiently pointed out to one of these persistent inquisitors that the answer to his question was thoroughly treated in one of his own papers!

After each lecture some of the participants joined Wiles for tea in the Fine Hall Common Room where discussion continued on the material of his lecture and on subjects related to the proof. There Wiles, always in good spirits, would continue to carefully answer questions and to proffer commentary.

I was present at all of those teas, and never once did I observe somebody ask Wiles how he was progressing on closing the gap, even though it was a question foremost in everyone's mind. It was, however, a question frequently asked of Wiles's students, Conrad and Skinner, but they always politely declined to comment.

John Conway made the following observation:

> Well you know we were behaving a little like Kremlinologists. Nobody actually liked to come out and ask him how he was getting on with the proof so somebody would say, "I saw Andrew this morning." Did he smile? "Well yes, but he didn't look too happy."[24]

Fred Diamond, Henri Darmon, Jerry Tunnell, Larry Washington, Shou-Wu Zhang, Brian Conrad and Chris Skinner were always available to answer technical questions, and they gave freely of their time. Washington gave to a few of us a preprint of his paper, *"Wiles' Strategy,"* and Darmon distributed to anybody who asked for it a preprint of his sixty-three page monograph, *"The Shimura-Taniyama Conjecture (D'Apres Wiles)."* Conrad distributed a fifty page manuscript entitled, *"Filtered Modules, Galois Representations, and Big Rings"* and a fifteen page manuscript entitled, *"Assorted Extras and Tidbits."* Skinner would make copies of the notes, from which he lectured, for anybody who asked for them. All of the aforementioned material was extremely helpful.

Richard Taylor was in residence during the entire spring term, and he gave a colloquium lecture in Taplin Auditorium early in the term. It was not generally known outside of Princeton that he was at that time collaborating with Wiles to close the gap in his proof, but he and Wiles were observed on occasion talking mathematics

1. February 10, 1994

together in the Fine Hall Common Room so it was generally assumed at Princeton that they were working on closing it together.

Periodically The Institute for Advanced Study asks prominent mathematicians to deliver an Institute Lecture on an important topic but directed to the layperson. On May 6th at 4:30 PM in Wolfensohn Hall, Wiles delivered such a lecture entitled, "Elliptic Curves" (cf. Photograph 25). I counted four Fields-Medal-recipient "laypersons" in the audience.[25]

Although his lecture was necessarily very elementary, he did touch upon the gap in his proof, and at the very end of his lecture he said that it was really inconceivable to him that his criterion that had already successfully established that there is an infinity of modular semistable elliptic curves with distinct j-invariants would fail to yield the general, semistable case. He said that all known results and conjectures in the field pointed to its correctness and that he was confident that he would ultimately be successful in his attempt to resolve the thorny, relevant technicalities.

A reception was held in the Fuld Hall Common Room after his talk. The next day Wiles developed a severe case of the flu, and within a day or so, his wife gave birth to their third daughter.

The following is a representative selection of the commentary at the time, both pro and con.

In the March 1994 issue of the *Notices of the American Mathematical Society,* Allyn Jackson wrote a two page article entitled, "Update on Proof of Fermat's Last Theorem." The main thrust of the article was that, as previously pointed out by Peter Sarnak, even if Wiles did not close the gap in his proof, his result that there exists an infinity of semistable modular elliptic curves over **Q** each associated with a distinct j-invariant constituted a fundamental contribution to mathematics.

In the March 1994 issue of *Scientific American* there appeared a short, uncredited commentary entitled, "Fermat's Theorem Fights Back." In the article appears the following:

> The very fact that Wiles is so competent, Faltings points out, means that he must be facing an extremely difficult and perhaps insurmountable problem. "If it were easy, he would have solved it by now," says Faltings, whose work helped Wiles to construct his proof. "Strictly speaking," Faltings comments, Wiles's recent travails suggest that "it wasn't a proof when it was announced."

In the June 28, 1994 edition of the *New York Times* Gina Kolata wrote a short article entitled, "A Year Later, Snag Persists In Math Proof." In the article appears the following:

> Dr. André Weil (pronounced VAY), a distinguished elder mathematician at the Institute for Advanced Study in Princeton, is a doubter. "I am willing to believe he has had some good ideas in trying to construct the proof, but the proof is not there," he said in an interview in *Scientific American*. "Also, to some extent, proving Fermat's theorem is like climbing Everest. If a man wants to climb Everest and falls short of it by 100 yards, he has not climbed Everest."

1. February 10, 1994

1. Andrew Wiles, Isaac Newton Institute for Mathematical Sciences, Cambridge University, Cambridge, England, June 23, 1993.

2. Andrew Wiles explaining where the gap has occurred in his proof, Taplin Auditorium, Fine Hall, February 10, 1994.

3. Nicholas Katz at the Goldfeld-Jacquet-Zhang Seminar, Columbia University, February 9, 1998.

1. February 10, 1994

4. Barry Mazur, Fine Hall, March 10, 1994.

1. February 10, 1994

5. Jean-Marc Fontaine, Conference in Honor of Barry Mazur, Science Center A, Harvard University, May 30, 1998.

6. Peter Sarnak at the Goldfeld-Jacquet-Zhang Seminar, Columbia University, April 29, 1997.

1. February 10, 1994

7. John Coates, Columbia University, April 22, 1999.

8. Jean-Pierre Serre, Conference in Honor of Barry Mazur, Science Center A, Harvard University, May 27, 1998.

1. February 10, 1994

9. Jean-Pierre Serre and John Tate, Conference in Honor of Barry Mazur, Faculty Club, Harvard University, May 27, 1998.

10. Gerd Faltings at the 250th Anniversary Celebration of the Founding of Princeton University, Taplin Auditorium, Fine Hall, March 19, 1996.

1. February 10, 1994

11. Andrew Wiles and Gerd Faltings at the 250th Anniversary Celebration of the Founding of Princeton University, Taplin Auditorium, Fine Hall, March 19, 1996.

12. Pierre Deligne with Phillip Griffiths at right, Wolfensohn Hall, Institute for Advanced Study, April 1993.

1. February 10, 1994

13. Henri Darmon, Fermat Conference, Jacob Sleeper Hall, Boston University, August 18, 1995.

1. February 10, 1994

14. Fred Diamond, Fermat Conference, Jacob Sleeper Hall, Boston University, August 18, 1995.

1. February 10, 1994

15. Larry Washington, Fermat Conference, Jacob Sleeper Hall, Boston University, August 18, 1995.

16. Shou-Wu Zhang, Columbia University, October 14, 1996.

1. February 10, 1994

17. Lucien Szpiro, Mathematics Building, Columbia University, November 16, 1998.

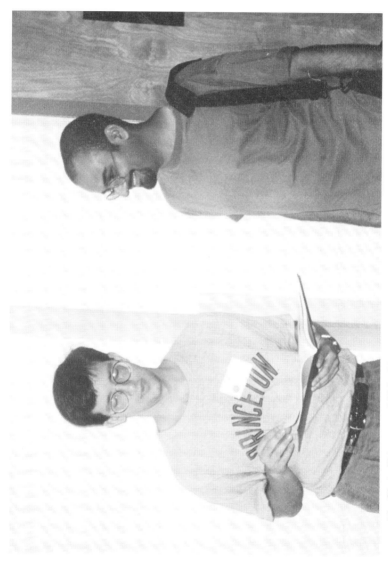

18. Chris Skinner and Nike Vatsal, Fermat Conference, Boston University, August 1995.

1. February 10, 1994

19. Brian Conrad, Fermat Conference, Jacob Sleeper Hall, August 15, 1995.

20. Jerry Tunnell, Fermat Conference, Jacob Sleeper Hall, August 1995.

21. Hervé Jacquet, Hill Center, Rutgers University, February 1993.

22. Ilya Piatetski-Shapiro, Fuld Hall, Institute for Advanced Study, November 18, 1999.

1. February 10, 1994

23. Joseph A. Shalika, Krieger Hall, The Johns Hopkins University. November 2, 1998.

24. Robert Langlands, Fuld Hall, Institute for Advanced Study, October 25, 1990.

1. February 10, 1994

25. Andrew Wiles, Wolfensohn Hall, Institute for Advanced Study, May 6, 1994.

Chapter 2

September 19, 1994

In the fall term of 1994 Jim Arthur delivered twice per week the Hermann Weyl lectures at the Institute for Advanced Study. His first lecture was on October 11th. Wiles was faithfully attending these lectures when suddenly he stopped attending. I noticed this, and on October 26th at the end of Arthur's lecture I approached a member of Wiles's inner circle, and I asked him if Wiles's absence meant that we would soon be getting another announcement from him. The strange look on his face and the fact that he immediately changed the subject indicated to me that something was about to break. I felt that it would be either an announcement that Wiles has closed the gap in his proof or an announcement that he was releasing his manuscript to allow others to work on closing the gap.

So I was not too surprised when the next day I read the headlines of a two column article that started in the upper left-hand corner of the front page of the *New York Times*, "Theory of Universe's Age Poses New Cosmic Puzzle... While a Mathematician Calls Classic Riddle Solved." The article was continued to page B12 where the section associated with Wiles's work occupied approximately one-half of the page. In the middle of the article there was a large photograph of Wiles sitting at his desk in his, by now, legendary attic office in his Princeton home. The main point of the article was that the gap in the proof had been jointly closed by Wiles and Richard Lawrence Taylor, who was associated with Cambridge University in England. The article failed to mention that Taylor was a former Ph.D. student of Wiles.

The following is the e-mail that Karl Rubin sent to the mathematics community:

> Subject: Update on Fermat's Last Theorem
> Date: 25 Oct 1994 11:04:11
>
> As of this morning two manuscripts have been released
>
> Modular elliptic curves and Fermat's Last Theorem by Andrew Wiles
>
> Ring theoretic properties of certain Hecke algebras by Richard Taylor and Andrew Wiles
>
> The first one (long) announces a proof of, among other things, Fermat's Last Theorem, relying on the second one (short) for one crucial step.
>
> As most of you know, the argument described by Wiles in his Cambridge lectures turned out to have a serious gap, namely the construction of an Euler system. After trying unsuccessfully to repair the construction, Wiles went back to a different approach, which he had tried earlier but abandoned in favor of the Euler system idea. He was able to complete his proof, under the hypothesis that certain Hecke algebras are local complete intersections. This and the rest of the ideas described in Wiles's Cambridge lectures are written up in the first manuscript. Jointly,

2. September 19, 1994

Taylor and Wiles establish the necessary property of the Hecke algebras in the second paper.

The overall outline of the argument is similar to the one Wiles described in Cambridge. The new approach turns out to be significantly simpler and shorter than the original one, because of the removal of the Euler system. (In fact, after seeing these manuscripts Faltings has apparently come up with a further significant simplification of that part of the argument.)

Versions of these manuscripts have been in the hands of a small number of people for (in some cases) a few weeks. While it is wise to be cautious for a little while longer, there is certainly reason for optimism.

Karl Rubin
Ohio State University

Wiles explained it this way:

On Monday, September 19th I was sitting here at this desk [cf. Photograph 40] when suddenly, totally unexpectedly, I had this incredible revelation. It was the most, the most important moment of my working life. Nothing I ever do again will [at this point Wiles paused to compose himself]... I'm sorry. I decided to go back and look one more time at the original structure of Flach and Kolyvagin to try and pinpoint exactly why it wasn't working, to try and formulate it precisely. One can never really do that in mathematics, but I just wanted to set my mind at rest that it really couldn't be made to work. I was trying, convincing myself that it didn't work, just seeing exactly what the problem was when suddenly, totally unexpectedly, I had this incredible revelation. I realized what was holding me up was exactly what would resolve the problem I had in my Iwasawa theory attempt three years earlier. It was the most, the most important moment of my working life. It was so indescribably beautiful, it was so simple and so elegant, and I just stared in disbelief for twenty minutes. Then during the day I walked around the department. I would keep coming back to my desk and looking to see if it was still there. It was still there. Almost what seemed to be stopping the method of Flach and Kolyvagin was exactly what would make horizontally Iwasawa theory, my original approach to the problem from three years before, work. Out of the ashes seemed to rise the true answer to the problem. So the first night I went back and slept on it. I checked through it again the next morning, and by 11 o'clock I was satisfied, and I went down and told my wife. "I have got it, I think I got it, I have found it." It was so unexpected, I think she thought I was talking about a children's toy or something, and she said, "Got what?" And I said, "I have fixed my proof, I have got it."

My wife has only known me while I have been working on Fermat. I told her a few days after we got married. I decided that I really only had time for my problem and my family, and when I was concentrating very hard, I found that with young children that is the best possible way to relax. When you're talking to young children, they simply are not interested in Fermat, at least at this age, they want to hear a children's story, and they are not going to let you do anything else.[1]

2. September 19, 1994

In his well-known book (see [**46**]), reviewed by no less than G. H. Hardy, *The Psychology of Invention in the Mathematical Field*, Jacques Hadamard enumerates the four stages of the creative process: (1) preparation, (2) incubation, (3) illumination, (4) verification. I can think of no more-dramatic example of stage (3) than what Wiles experienced on September 19, 1994.

On page 55 of his book Hadamard states:

> ...one rule proves evidently useful: that is, after working on a subject and seeing no further advance seems possible, to drop it and try something else, but to do so provisionally, intending to resume it after an interval of some months.

The Iwasawa-Flach-Iwasawa switching by Wiles contributes to establishing the validity of Hadamard's rule.

Wiles decided to give a very detailed, technical lecture restricted to the portion of his proof that required correction, and a very poorly announced lecture was scheduled for Monday, October 31st in Taplin Auditorium. There were approximately thirty-five people present when this historic event occurred. I was very fortunate in that the night before Justine Bumby, the mother of John Grothendieck, called me and made me aware of the lecture.

At the beginning of the lecture Wiles stated that Gerd Faltings had carefully reviewed his joint work with Taylor and not only decided that it was correct but also was able to somewhat simplify the proof. Never once did he mention or allude to Fermat's Last Theorem. Photographs 27 through 30 are in time sequence of Wiles during that lecture.

The next day in my exuberance I brought my very crude tape recording of the lecture to a professional sound laboratory to have it digitally enhanced and duplicated, but when Andrew asked me not to distribute it, I gave the copies to him.

There was a champagne reception immediately after the talk in the Fine Hall Common Room. Wiles's wife, Nada, and infant daughter (asleep in a carriage) as well as his two young daughters were in attendance. I was talking with her, when Joe Kohn came over and offered to her his congratulations. In the course of the conversation he pointed out that once an error is found in an attempted proof of a very important, unsolved problem, it is very rare that the author is able to recover and ultimately establish the result. There was a table on which there had been placed a large pile of preprints of the Wiles and the Wiles-Taylor manuscripts available to anybody who wanted to take them.

On December 16, 1995 when I was visiting Peter Sarnak at his Princeton home and we were discussing the Wiles-Fermat saga, he told me that Faltings had performed a very great service for mathematics in October of 1994 by placing his reputation on the line and pronouncing the Wiles-Taylor work to be correct. With regard to Fermat, mathematics was in a state of agitated turmoil at that time, and by his courageous action he put an immediate end to all of the confusion, speculation and controversy. It required a man of his respected reputation and integrity to accomplish what he had done.

Faltings explained to me his involvement:

> Thank you for your letter. My role in the Fermat business can be described as follows: (1) In the initial try (1993) Wiles kept me totally uninformed and also did not show the manuscript to me. (2) In the summer of 1994 I looked a little bit into the matter using notes which were semipublic at that time. This was because I was running a summer school for students, and in the announcement (which had been formulated in the summer of 1993) I had promised

2. September 19, 1994

to explain Wiles's method expecting it to be public knowledge at the time of the summer school. I did not find the famous "mistake," but I obtained some familiarity with the methods. (3) In October of 1994 Wiles sent to me a manuscript with the new proof asking me to check it. This I did, and Wiles went public. I also managed to find some simplifications on the way, but my role was no more than that of a referee in the usual process of publishing papers.[2]

It should be pointed out that although it may not have had direct application to Fermat, some of Faltings's previous work is incorporated into Wiles's Annals paper. Also, in 1983 Faltings proved, among other things, Mordell's conjecture, which immediately implies that if Fermat's Last Theorem were false for a particular prime exponent, then there could be at most a finite number of solutions in the positive integers with respect to that exponent. This was the deepest known result on Fermat until Wiles, himself, published a proof.[3]

Sarnak also told me that during December of 1993 he persuaded Wiles to collaborate with an expert, whom he could trust, in the area of Hecke algebras in his effort to close the gap in the proof.

On a future occasion Sarnak would say:

> Every time he would try and fix it in one corner it would produce some other difficulty in another corner. It was like he was trying to put a carpet in a room where the carpet had more area than the room, but he could put it in any corner and then when he ran to the other corner it would pop up in the original corner, and whether you could put the carpet in the room was not something that he was able to decide.[4]

2. September 19, 1994

The October 31st edition of *The Daily Princetonian*, the university student newspaper, had a story about Wiles's achievement that started in the upper right-hand corner of the front page and continued to the entire left column on page 3. The headline read, "Wiles Reveals New Proof of Theorem." There was a headshot of Wiles with the caption, "May have filled gap in proof." The next day I went to the office of the newspaper, and the editor kindly gave me approximately one hundred copies of that issue.

Wiles gave lectures to his graduate class in room 214 Fine Hall, in which he further elaborated on the work that he and Taylor had done as well as on other aspects of his proof, on November 2nd, 9th, 16th and 23rd and on December 7th. In the same room Conrad gave lectures on November 7th, 14th, 21st, 28th and on December 12th and 14th. It seemed odd to me that after it was known that Wiles had corrected the proof, the attendance at his lectures and at those of his graduate students decreased vis-à-vis before he corrected the proof.

In Taplin Auditorium at 4:30 PM on November 30th, Wiles gave a technical lecture primarily on the work that he and Taylor had done (cf. Photograph 32). This talk was announced several weeks in advance, every seat was taken, and people were standing along the rear wall. When he finished his talk, Joe Kohn presented Wiles with a large cake on behalf of the Department.

That evening there was a dinner in his honor in the Institute dining room. Wiles gave a brief speech just before the dinner started in which he said, among other things, that he had recently received an e-mail from Fermat... from heaven... requesting a copy of his manuscript. I had the pleasure of sitting with my friend, John Nash, at the dinner, who the next morning left for Stockholm to receive the Nobel Prize.[5]

On December 3rd there was a nice party at the lovely home of Robert Langlands for Jim Arthur and those who attended his lectures during the term. I sat with Peter Sarnak and his wife and Robert

2. September 19, 1994

MacPherson. Surprisingly, the conversation did not focus on Wiles-Fermat. However, after the dinner I had a brief conversation with Langlands on that subject, and he informed me that the 1995–96 academic year at the Institute would be devoted to Wiles's proof and many experts in his field would be in attendance. In fact he said that this had been decided before Wiles announced that the gap had been closed. I was impressed with his modesty vis-à-vis the Langlands-Tunnell result, which is crucial to the proof.

On December 5th Wiles gave a lecture on his proof at Harvard University in Science Center B, a very large auditorium, which was filled to capacity but which unfortunately had a less than satisfactory sound system. Mathematicians were present from many institutions. Barry Mazur was his host and, when he introduced Wiles, welcomed the audience to Harvard's "Basic Notions" seminar.

This was Wiles's first lecture on his proof outside of Princeton since his lectures at Cambridge University in June of 1993, and there was an atmosphere of great excitement before and during the lecture. A Japanese gentleman sitting to my right was taking photographs with what must have been one of the first commercially available digital cameras.

Wiles seemed to devote an inordinate amount of time to discussing various equivalent formulations of the Shimura-Taniyama conjecture so that he had to rush as he approached the end of his lecture to cover all areas of the proof. When he finished, he turned to the audience with a broad grin (cf. Photograph 33) and received a prolonged and very enthusiastic ovation. These are the moments in life that one never forgets. I observed more than one person in the audience with tears in their eyes.

On December 10th Wiles delivered the Dolciani Lectures at Hunter College in room 714 in the West Building.[6] He spoke in two sessions: 11:00 AM-12:00 PM and 1:30 PM-2:30 PM. He was careful to mention that his host, Joseph Roitberg, had invited him to speak before the gap in his proof had been closed (cf. Photograph 34).

There was a very animated question and answer session after the second lecture.

Wiles pointed out that although he had a fascination with Fermat's Last Theorem in his youth, until it was linked by Ribet to modern, mainstream mathematics he did not pursue it during his professional career as a research mathematician. The fact that Ribet had linked Fermat to Shimura-Taniyama provided him with further motivation to attack the problem since even if he did not solve it completely, there was the real possibility that he would produce something of genuine mathematical interest toward the solution of the latter conjecture.

When asked what he thought was the next major unsolved problem that mathematicians should try to solve, he said without hesitation that most mathematicians, himself included, would say the Riemann hypothesis.[7] However, he very characteristically did not give even the slightest indication as to whether or not he was seriously working on that problem.

In response to the question as to whether or not Fermat had a proof (a question that was asked at virtually every lecture he gave) Wiles said that Fermat was in the habit of challenging mathematicians, especially English mathematicians, with problems that he had solved and the fact that he never challenged the English mathematicians with his Last Theorem indicates to him that Fermat probably had ultimately realized that he did not possess a proof after all.

On the front page of the Science Times section of the January 31, 1995 edition of the *New York Times* there appeared a story by Gina Kolata entitled, "How a Gap in the Fermat Proof Was Bridged," devoted to the circumstances surrounding the closing of the gap in Wiles's proof. The article occupies the bottom half of that page, and it is continued to page C9 where it occupies approximately half of that page also. The article features a separate photograph of Wiles and a separate photograph of Taylor.

2. September 19, 1994

The *New York Times* had wanted a recent photograph of Wiles for the article and in fact was prepared to dispatch a staff photographer to Princeton to obtain one. Andrew sent me an e-mail in which he asked if I would provide a recent photograph. I was delighted and honored to do so and immediately sent to the *New York Times* photography editor a photograph that I took at Wiles's Princeton lecture on November 30th, for which the *New York Times* paid me the magnificent sum of one-hundred fifty dollars.

This article is so informative concerning the Wiles-Taylor collaboration and so well written that I have decided to quote significant excerpts from it:

> The enormous intellectual trophy of having conquered the world's most famous mathematical problem seemed about to slip from his grasp. If he invited a well-known mathematician to help him bridge the gap, he would risk having to share the credit. What was needed, and what he pulled off, was a miraculous save with just the right collaborator.
>
> "It's a very competitive environment," Dr. Wiles explained, and, after working for seven years and after getting so close, he wanted the victory to be his alone.
>
> **The Gap**
> **A 'Minor' Problem Turns Into a Crisis**
>
> The gap at first seemed to be a minor one. But though the estimate seemed intuitively to be correct, proving it was a different matter.
>
> "At first I thought I was just being slow," he said. "It can be very hard to tell at first whether you're just being blind to something or whether there really is an obstacle; it's very hard to tell the difference."

After struggling in vain, he decided to go into complete seclusion. "If I was to focus on the problem, I could not allow myself to be distracted," he said. He tried out ideas on a Princeton colleague, Dr. Nicholas Katz, but told no one else the details of what he was doing.

In the outside world, Dr. Wiles's secretiveness irritated many and touched off a betting game on what he was up to and whether he would succeed.

"I noticed he was stuck and I was very pessimistic," said Dr. Gerd Faltings, a mathematician at the Max Planck Institute in Bonn, Germany.

Dr. Kenneth Ribet of the University of California at Berkeley said that although he was rooting for Dr. Wiles, as time went by the unfilled gap in the proof was becoming something of an embarrassment. "A proof that is unfinished is no proof at all," Dr. Ribet said. "I had heard from people in other parts of mathematics that this whole algebraic geometry group has egg on its face, that we can't get our act together. They would say that we don't know a proof when we see one, so how could anyone believe what we say."

Unfortunately, Dr. Wiles said, the gap had turned up in the very part of the proof where he felt he was on the shakiest ground. He had used a new and powerful method devised by Matthias Flach of Princeton University. But, Dr. Wiles said, Dr. Flach's method, "used a lot of sophisticated machinery, and I didn't feel completely comfortable with it."

As months went by, Dr. Wiles finally decided that he needed help. He was frightened that he would

2. September 19, 1994

mislead himself, spending years, or decades, going in circles trying frantically to fix the gap with methods that never would work. "There are people who have spent 30 years on one problem," he said. "You get caught in problems, trapped in them. I knew the danger, psychologically."

And it was getting harder and harder to think clearly. "I was very tired," he said. "I'd been working very hard, and I needed someone to check every statement I made. I needed someone to talk to all the time." But Dr. Wiles needed a collaborator who would allow him to call the shots. "I wanted to choose the direction—I didn't want someone to come in with lots of new methods," Dr. Wiles said. Before he gave up and let the world in on the problem, he said, "I wanted to make sure that I had completely explored the ideas that I had and that I hadn't missed something obvious."

The glory of being the person to have proved the theorem was too overwhelming to give up without a fight. Dr. Wiles quoted Enrico Bombieri, a mathematician at the Institute for Advanced Study in Princeton, who said: "Everyone has their price. For mathematicians, it's Fermat's Last Theorem or the Riemann hypothesis." Riemann is another outstanding unsolved problem.

Dr. Wiles thought carefully about whom he would ask to join him. "I wanted someone I was sure of," he said. So he called Dr. Richard Taylor, 32, a former student who is now a reader, a tenured position, at Cambridge University.

Dr. Taylor said he was surprised and excited when Dr. Wiles called him in December 1993 to ask for help. He had a sabbatical coming up, so he arrived in Princeton last January ready to work intensively on the problem. While Dr. Wiles had "extremely good" intuition, Dr. Taylor said, "I'm somebody who is more concerned with details."

Dr. Wiles and Dr. Taylor spent several months trying to see whether Dr. Flach's methods could be made to work. Then, over the summer, they tried a different approach, one that was a long shot, in Dr. Wiles's opinion. Finally, Dr. Taylor suggested that they go back to Dr. Flach's method once again. Dr. Wiles was reluctant. "I confess that at that time I was thinking that it would need a whole new approach that would take several years and require many people," he said.

But he went back to the Flach method for one last time. "There was one variant in the original argument that I'd convinced myself wouldn't work but I hadn't convinced him," Dr. Wiles said. "I was sitting at my desk one morning really trying to pin down why the Flach method wasn't working when, in a flash, I saw that what was making it not work was exactly what would make a method I'd tried three years before work. It was totally unexpected. I didn't quite believe it." He dashed down from the attic to tell his wife. Although his enthusiasm was infectious, Dr. Wiles said, "I actually think she didn't believe me."

Dr. Wiles tried to control his mounting elation. "I was too excited to think clearly," he said. "I slept on it," he said, and in the morning he called Dr. Taylor who, by this time had returned to England. The two began collaborating furiously and, two and

2. September 19, 1994

a half weeks later, they had written a paper, whose authors are Dr. Wiles and Dr. Taylor, that filled the gap of the proof.

The credit for solving Fermat's Last Theorem seems likely to go largely to Dr. Wiles, since he is the sole author of the major paper on the proof, and since it was his ideas that bridged the gap. Although Dr. Taylor evidently played an important supporting role, he says that he agrees with Dr. Wiles's account of their collaboration. Dr. Diamond carefully describes the proof as "the theorem of Wiles, completed by Taylor and Wiles." Dr. Wiles said he agreed with this description.

The Breakthrough
Discovering Gold in 'Pigeonholes'

Dr. Wiles said that the breakthrough came in figuring out how to glue together an infinite collection of mathematical objects called Hecke rings. He had initially been creating what was a "very natural relationship between these objects, natural in the sense that you can give a clear definition of the maps between them." It was an inductive argument. The idea was to take one element of a set and use that to find the next element, then to use the second to construct a third, and so on.

The new idea, Dr. Wiles said, "was to simply construct artificial maps between these objects."

"You wouldn't show a relationship explicitly," he said, but would use a counting argument to prove that a relationship had to exist. The basic idea is to use the pigeonhole principle: if you have more objects than pigeonholes to put them in, then at least one pigeonhole must contain more than one object.

The complete argument involves creating an infinite sequence of sets of pigeonholes and then showing that there must be objects that show up in every set of pigeonholes. This allowed Dr. Wiles and Dr. Taylor to prove that there must be an infinite set of Hecke rings that share a relationship, although they never have to specify exactly what that relationship is.

Before Dr. Wiles was willing to announce that he and Dr. Taylor had filled the gap, he asked a few leading experts to check his argument. One was Dr. Faltings in Germany, who said he read it in a week and was convinced it was correct. Now, Dr. Faltings has improved the proof, making it sleeker and easier to follow. Still, he said, most mathematicians who are expert enough in the field to read the proof will probably require a month to go through it.

Another who got the proof early was Dr. Diamond, 30. "I certainly didn't expect that kind of argument," Dr. Diamond said. "It's so ingenious."

A few weeks later, he discovered how to extend it to prove a larger section of the Taniyama conjecture. The conjecture involves elliptic curves defined over the rational numbers. Dr. Wiles and Dr. Taylor showed it was true for a subset of elliptic curves, including those known as semistable curves. Dr. Diamond showed it was true for a much larger group of elliptic curves.

2. September 19, 1994

The Next Frontier
On to Taniyama; Then Langlands?

Now, mathematicians said, the final proof of the elusive Taniyama conjecture is in sight. "If I had to guess, I would say it is at least a few months and at most a few years," before the entire conjecture is proved, Dr. Diamond said.

And, finally, the work is a major step in the Langlands program, the grand unified theory of mathematics. Dr. James Arthur of the Institute for Advanced Study in Princeton said the Langlands program, a conjecture proposed a quarter-century ago by Dr. Robert Langlands, united two seemingly disparate fields of mathematics. It says that the mathematics of algebra, which involves equations, and the mathematics of analysis, which involves the study of smooth curves and continuous variations, are intimately related.

The conjectures in the Langlands program "are like a cathedral, the way they fit together so beautifully," Dr. Arthur said. Yet, he said, "they are so difficult to prove."

"The only way to prove them seems to be by very, very indirect means," he said, adding that he thought it would take "decades at the very least, maybe centuries," to complete the Langlands program.

Nonetheless, Dr. Arthur said, the greatest advance so far has been the work of Dr. Wiles and Dr. Taylor and the recent result by Dr. Diamond. "It was an enormous accomplishment, what Wiles has done, really a huge step," Dr. Arthur said.

Dr. Wiles does not intend to work on the Langlands program, and he has no intention of extending his proof or mining it for new results. "It is nice to finish it off, but, for me, the mystery is gone," he said. "The challenge now is to go on to problems where no one has any idea where to start."

During the spring term, again in room 214, Conrad lectured on February 6th, 8th, 20th, 22nd, 27th and on March 1st and 6th. Skinner lectured on March 20th, 27th, 29th and April 3rd, 5th, 10th, 12th, 17th, 19th, and 26th. They were in their very early twenties, and they did a most impressive job. I am not at all surprised that at this time Skinner is completing a rarely-given three year appointment at the Institute for Advanced Study, and Conrad is on the faculty of Harvard University.

After the gap in the proof was closed, Wiles was inundated with requests for interviews and lectures. One interview request to which he devoted significant attention was that of Ian Katz, a science writer for the British newspaper, *The Guardian*. His subsequent article (entitled: Fame by Numbers) appeared in the Weekend magazine section of the April 8, 1995 edition of that paper, and it occupies almost six full pages. It contains three very compelling color photographs, one each of Wiles, Ribet and Faltings.

This article is also so informative and so well written that I have decided to quote significant excerpts from it:

> He is grinning now as he talks about his beloved proof. He's extolling its "feeling of naturalness and elegance" his long fingers reaching out in front of him as though caressing his mathematical edifice. What gives it its beauty, he is explaining, is that its "deep results come not with thousands and thousands of careful calculations but from real ideas that haven't

2. September 19, 1994

been around before. People haven't looked at something in the right way and suddenly everything comes together and there's a simplicity and clarity that's very beautiful."

His father, the venerated theologian Maurice Wiles, would later become Regius Professor of Theology at Oxford. His mother taught mathematics. His older brother is a university lecturer too; he teaches drama.

According to his sister Alix, a psychologist by training, Wiles shone academically from the start. "My eldest brother was very clever. He always got very glowing reports, but Andrew's were more glowing." Though he was "on the shy side," he wasn't nerdish: he played cricket and tennis for his school—The Lees in Cambridge—and insisted on doing English A-Level to break up the standard scientist's quartet of double mathematics, physics and chemistry. Alix believes that breadth of interest was a factor in his later success: "It enabled him to pull together fields and open up things that other people couldn't see." After A-Levels—in which he coyly recalls earning "four As, I guess"—Wiles went to Merton College, Oxford, to study mathematics. His sister remembers him having a "pretty grim time" there, but Wiles himself goes to some lengths to "correct" her version, insisting that he enjoyed himself and that he maintains close links with the college. Whichever is closer to the truth, he surprised no one by collecting a First and moving back to Cambridge, the real center of British mathematics, to study for a Ph.D. He arrived in Cambridge at a time when two British mathematicians, Peter Swinnerton-Dyer and Bryan Birch, had re-awakened interest in the study of so-called "elliptic curves," a field which had been out of fashion for most of the century. The curves, which can be described by the general formula $y^2 = x^3 + ax + b$, fascinate mathematicians because they exhibit a complex structure which yields insights in many different fields. For his Ph.D. thesis, Wiles attacked the conjectures of Swinnerton-Dyer and Birch head-on, making a series of theoretical breakthroughs, sometimes in collaboration with his Australian-born thesis supervisor John Coates. "These problems were 100 years old and many people had worked on them, but he saw right through them," Coates recalls.

He was offered a research fellowship at Clare College but instead slipped down the brain drain to Harvard where he spent three-and-a-half years. After shorter stints in Paris, Bonn and at Princeton's Institute for Advanced Study, he accepted a full professorship in the Ivy League college's high-octane mathematics department in the early Eighties.

By then a string of impressive results, including a part in proving the important so-called Main Conjecture of Iwasawa, had earned him a reputation as one of the most rigorous and deep thinkers in the field of number theory. But he was by no means the top dog. That distinction went unequivocally to Gerd Faltings, a brilliant—and brusque—young German who for several years was a Princeton colleague of Wiles. By his mid-thirties, Faltings had revolutionalized the field of number theory, collected a Fields medal—the mathematical equivalent of a Nobel prize—and, some say, developed an ego to match. "There was a profound rivalry between them when Faltings was at

2. September 19, 1994

Princeton," recalls Alix. "Before Fermat they were always one-upping each other and Faltings always won."

Wiles insists that he was unembarrassed by the discovery of the flaw. "Why should I feel embarrassed? I'd made a sensational breakthrough, even without this. I'd opened the door on the subject. Even if I'd never fixed it I'd made an enormous advance." "Most people don't have the guts to work on something where they don't know where to begin. But I'd removed the mystery from this problem. Now, of course, they [other mathematicians] could see the end was in sight." A delay of a month or two while Wiles tried to rescue the proof himself might have been fair enough, says Faltings, "but he did this for too long." The German was not the only mathematician becoming annoyed by the Englishman's secretiveness. At Princeton, Wiles's colleagues were fielding an increasingly irritable torrent of enquiries about when the proof would be made available. Wiles had long since stopped responding to e-mail.

Wiles had taken the decision to isolate himself from the outside world. "People would say you know there's a real storm out there and I'd say, 'I don't want to know about it'." But his mood was deteriorating. He began to worry that he might never crack the problem, that he might be "caught in a vortex" like other mathematicians he knew about. "Once you were caught you could spend years and years in this intense and unresolved state."

At the same time he was becoming more and more convinced other mathematicians were working on the problem. At Harvard, Barry Mazur was running a program on the subject. Ribet was lecturing on it, though he says he decided not to work the copy of Wiles's original manuscript he had been sent to referee. Wiles's biggest fear, however, was Faltings. "Andrew believed from the beginning that he was hard at it," recalls his sister Alix.

There is one niggling question that hovers, like an unfinished crossword, over this whole saga. Did Fermat really have the "truly marvelous demonstration" that would not fit into the margin of his book? Was he little more than a kind of mathematical hellraiser who enjoyed torturing his rivals with inflated claims? Or did he really have an elegant, simple proof that could make even Wiles's spectacular edifice look flabby and overworked?

For at least a century most mathematicians have inclined towards a third possibility: that he thought he had discovered a proof but had been mistaken. In his account of Fermat's career Mahoney suggests that the French mathematician believed a technique known as his "method of infinite descent" could be adapted to prove his Last Theorem. Like the supposed proof of his Last Theorem, he had written that he could not fit a full description of his method of infinite descent into the margin. But mathematicians have struggled in vain to tease a proof of the Last Theorem from the "descent" idea.

For his part, Wiles insists that he is not troubled by the idea that a simple solution lies waiting to humiliate generations of mathematicians. "We understand Fermat's methods so well now that we can safely assume that he didn't find some completely

2. September 19, 1994

new approach to it. If he had a proof then it must have been a very, very cunning use of methods we know well, and the idea that only he could see it with methods known and used for 200 years is pretty hard to accept."

There is, he admits, "a tiny doubt," but it's one he can live with. At last, he says, he feels he can relax. "It has set me free. I was caught up in this problem. It isn't just like conquering something outside me. It was a part of me."

On April 21st Wiles delivered a lecture on his proof in room 220 of the Durham Laboratory of Engineering at Yale University between 3:30 PM and 4:45 PM (cf. Photograph 35). During the course of the lecture he said that when he first started to work on Fermat, he casually mentioned that fact to a member of Yale's mathematics department, and based on that experience he came to the conclusion that it would be best that he not continue to tell people that he was working on the problem.

On a future occasion Wiles would explain it this way:

Certainly for the first several years I had no fear of competition. I simply didn't think I or any one else had any real idea how to do it, but I realized after a while that talking to people casually about Fermat was impossible because it just generates too much interest and you can't really focus yourself for years unless you have this kind of undivided concentration which too many spectators will have destroyed.[8]

On April 24th Wiles delivered a one hour lecture on his proof in room 1302, Weaver Hall, Courant Institute of Mathematical Sciences, New York University (cf. Photograph 36).

It was most interesting to observe six or seven of the world's foremost authorities on differential equations sitting together in the front row in rapt attention while one of the world's foremost authorities on algebraic geometry and homological algebra lectured to them on one of the most abstract proofs in all of mathematics. Interesting indeed!

On April 27th Wiles delivered a lecture on his proof in Schapiro Hall Auditorium at Columbia University (cf. Photograph 37). This very large room was filled to capacity with mathematicians from many institutions, and, as had occurred at Harvard, he received a prolonged and very enthusiastic ovation at the end of his talk.

Dorian Goldfeld, his host, had very carefully distributed announcements of the lecture to colleges and universities in the New York City area and posted an announcement on the Internet to afford to as many mathematicians in the area as possible notification of Wiles's intended lecture.

On May 15th Wiles delivered one of the finest general audience lectures that I have ever experienced (cf. Photograph 38). It was held in Harold Helm Auditorium in McCosh Hall at Princeton University. His wife, who holds a doctorate in biology, was present.

After Wiles was introduced by Joseph Kohn, he walked to the podium and said

> *Cubum autem in duos cubos, aut quadrato-quadratum in duos quadrato-quadratos, et generaliter nullam in infinitum ultra quadratum potestatem in duos ejusdem nominis fas est dividere; cujus rei demonstrationem mirabilem sane detexi. Hanc marginis exiguitas non caperet.*[9]

...on the other hand...

It was vintage Wiles.

2. September 19, 1994

This was the first time I became aware of the now oft-quoted statement that doing mathematics is analogous to finding one's way in a darkened room. On a future occasion Wiles would state it this way:

> Perhaps I could best describe my experience of doing mathematics in terms of entering a dark mansion. One goes into the first room and it's dark, completely dark. One stumbles around bumping into the furniture and then gradually you learn where each piece of furniture is, and finally after six months or so you find the light switch. You turn it on; suddenly it's all illuminated, you can see exactly where you are.[10]

In the July 1995 issue of the *Notices of the American Mathematical Society*, Gerd Faltings presented a four page sketch of the proof of Fermat's Last Theorem.

The preeminent French photographer, Henri Cartier-Bresson, espoused the concept of "the decisive moment" when all of the visual and the emotional elements come together to express the meaning of a scene. On October 22, 1987 I hosted at the Nassau Inn in Princeton a dinner party to celebrate the 80th birthday of S. Chowla when Photograph 26 of Shimura and Weil in animated conversation was taken at "the decisive moment."

During the fall term of the 1995–96 academic year, Jerry Tunnell devoted his graduate course at Rutgers University to the work of Ribet and Wiles and others needed to resolve Fermat's Last Theorem. He was especially careful to develop the background material necessary to properly understand their work as well as the research papers which support it. The course was very timely and well-received.

The following is the course description that Tunnell published on his web page:

> This course will cover topics in number theory which lead to the proof of Fermat's Last Theorem. Several themes drawn from number theory of the last 35 years will be considered, including those relating to elliptic curves, modular forms and representations of Galois groups. The starting point will be an overview of the proof to indicate the main ideas, followed by detailed examination of the relevant methods. Many of the most technical portions of the proof will be treated in a concurrent seminar. Topics will include some of the following:

1. Overview of the proof of Fermat's Last Theorem
    - Relation of integer solutions of $x^n + y^n = z^n$ to the existence of elliptic curves with special properties
    - Congruences for modular forms and Ribet's descent argument
    - Modular elliptic curves and Wiles's proof that sufficiently many elliptic curves are modular
2. Galois group representations attached to elliptic curves and modular forms
    - Deformations of modular forms and Galois representations
    - Ramification properties of Galois representations
3. Group cohomology and Selmer groups
4. Geometry and number theory of modular curves
5. Congruence properties of modular forms
6. Open problems in elliptic curves and Galois representations

> Prerequisites: For the first third of the semester only the usual first year graduate courses will be assumed.

2. September 19, 1994

Examination of the details of the proof will require knowledge of basic algebraic number theory and elliptic curves. At any point this knowledge could be replaced by a willing suspension of disbelief.

Course Format: There will be an optional seminar where members of the class will discuss details not covered in the main lectures. As usual in my graduate courses there will be assignments covering the methods introduced in lectures. No student will be required to do more than they are capable of.

On February 23, 1996 Wiles delivered a lecture devoted to the research of Tunnell concerning the classical diophantine problem to determine which integers are the area of some right triangle with rational sides (see [**123**]). He spoke at the DIMACS Center Auditorium, Rutgers University (cf. Photograph 39). Afterward a dinner was held at the Frog and Peach restaurant in New Brunswick, followed by a party in Wiles's honor at Tunnell's home.

Several people, when they learned about my photographs and about my writing a book, have asked me to comment on Wiles himself. So who is Andrew Wiles? Assuredly, like most highly intelligent and accomplished men he has a complex personality with many facets and with many dimensions, but on balance he is among the nicest people I have known. Polite, considerate, accommodating, affable and unassuming to a fault, it is almost impossible to not genuinely like him. I have endeavored to sprinkle samples of his dry, almost arid wit throughout the book, and I have whenever possible carefully chosen the photographs of him that most closely reveal the man as he appears to me. He is by nature quiet and somewhat shy and at times even a bit ethereal, but on numerous occasions I have observed him chatting amiably with mathematicians and nonmathematicians alike about the events of the day, and it is rare that he does not have a perceptive insight or two to share about them. He is a family man totally committed to his three children and to his wife. He is a teacher, who takes very seriously his obligations to the undergraduates at Princeton as well as to the more advanced graduate and postgraduate students. He is a scholar, who to this day with all of his fame and notoriety carefully and diligently takes notes whenever he attends a lecture or a seminar.

He was born April 11, 1953 in Cambridge, England. In 1971 he entered Merton College, Oxford University, and obtained the BA degree in 1974. He then went to Clare College at Cambridge University, receiving his Ph.D. in 1980.

From 1977 until 1980, he was a Junior Research Fellow at Clare College and a Benjamin Pierce Assistant Professor at Harvard University. In 1981 he was visiting professor at the Sonderforschungsbereich Theoretische Mathematik in Bonn. In the fall of that year he was a member of the Institute for Advanced Study. In 1982 he became a professor of mathematics at Princeton University and in the spring of that year he was a visiting professor at the University of Paris, Orsay. Supported by a Guggenheim Fellowship he was a visiting professor at the Institut des Hautes Etudes Scientifiques and at the Ecole Normale Superieure (1985–1986). From 1988 to 1990 he was a Royal Society Research Professor at Oxford University. In 1994 he was appointed Eugene Higgins Professor of Mathematics at Princeton University.

In 1989 he was elected a Fellow of the Royal Society, London. In 1995 he received the Schock Prize in Mathematics and the Prix Fermat.[11] In 1996 he shared the Wolf Prize with Robert Langlands, and he received the National Academy of Sciences Award in Mathematics.[12] In 1997 he received the Cole Prize and the Wolfskehl Prize, and he was named a MacArthur Fellow.[13] In 1998 he received the Faisal Prize and a special tribute from the International Mathematical Union.[14] In 2000 he was knighted "for services to science" by Queen Elizabeth II.

2. September 19, 1994

26. Goro Shimura and André Weil, Nassau Inn, Princeton, October 22, 1987.

27. Andrew Wiles, Taplin Auditorium, Fine Hall, October 31, 1994.

2. September 19, 1994

28. Andrew Wiles, Taplin Auditorium, Fine Hall, October 31, 1994.

29. Andrew Wiles, Taplin Auditorium, Fine Hall, October 31, 1994.

2. September 19, 1994

30. Andrew Wiles, Taplin Auditorium, Fine Hall, October 31, 1994.

31. Richard Taylor, Conference in Honor of Barry Mazur, Science Center A, Harvard University, May 1998.

2. September 19, 1994

32. Andrew Wiles, Taplin Auditorium, Fine Hall, November 30, 1994.

33. Andrew Wiles, Science Center B, Harvard University, December 5, 1994.

2. September 19, 1994

34. Andrew Wiles, West Building, Room 714, Hunter College, December 10, 1994.

35. Andrew Wiles, Durham Laboratory of Engineering, Room 220, Yale University, April 21, 1995.

2. September 19, 1994

36. Andrew Wiles, Warren Weaver Hall, Room 1302, Courant Institute of Mathematical Sciences, New York University, April 24, 1995.

37. Andrew Wiles, Schapiro Hall Auditorium, Columbia University, April 27, 1995.

2. September 19, 1994

38. Andrew Wiles, Harold Helm Auditorium, McCosh Hall, Princeton University, May 15, 1995.

39. Andrew Wiles, DIMACS Center Auditorium, Rutgers University, February 23, 1996.

2. September 19, 1994

40. Andrew Wiles in his office Room 805 Fine Hall, February 6, 1995.

41. The Princeton home of Andrew Wiles containing the attic office in which he worked, June 25, 1998.

Chapter 3

August 9, 1995

In 1995 from August 9th to August 18th the Instructional Conference on Number Theory and Arithmetic Geometry was held at the College of General Studies at Boston University. All lectures were given in Jacob Sleeper Hall, a very large auditorium in the college with an ample balcony.

The conference was advertised and promoted in the *Notices of the American Mathematical Society* as being conducted primarily for advanced graduate students and recent Ph.D. recipients to introduce them to the areas of mathematics associated with the Fermat proof.

The following is an excerpt from the Conference Announcement:

> Description: The conference will focus on two major topics: (1) Andrew Wiles's recent proof of the Taniyama-Shimura-Weil conjecture for semistable elliptic curves; and (2) the earlier works of Frey, Serre, and Ribet showing that Wiles's Theorem would complete the proof of Fermat's Last Theorem. In keeping with its "instructional" mission, the conference will begin with introductory lectures on elliptic curves,

modular curves, modular forms, and Galois representations. Wiles's work also draws from a significant number of more advanced topics, including the deformation theory of Galois representations, refined structure of Hecke algebras, complete intersection rings, and generalized Selmer groups. Each topic will be introduced by an expository lecture describing some of its history and explaining in general terms how it fits into the proof of Wiles's theorem. The ensuing lectures will cover the finer aspects of the proof in detail.

In recognition of the historical significance of Fermat's Last Theorem, some lectures will also reflect on the history of the problem while others may speculate on the future and describe some of the connections of Wiles's work with other parts of mathematics.

There were approximately four-hundred participants registered by the conclusion of the conference.

The conference was superbly organized and supervised by Glenn Stevens, Joseph Silverman, and Gary Cornell (cf. Photographs 53, 54, 58). It was easily the best mathematics conference I have ever attended.

The papers from the conference were published by Springer-Verlag in a single volume (see [**19**]) for which Ram Murty (cf. Photograph 62) has produced an extraordinary review, which is essentially a very-carefully-constructed outline of the proof itself (see Appendix D).[1] Also, a video tape recording was made of each of the lectures presented during the conference, and those recordings have been preserved in the Boston University Department of Mathematics.

I was especially impressed by the very high-level spirit of the conference. Everybody appeared to be excited about the results of Frey, Serre, Ribet, Wiles and Wiles-Taylor and the fact that one of

3. August 9, 1995

the grand unsolved problems of mathematics had been resolved. It is hard to describe the atmosphere at the conference, but I imagine anyone who was there knows exactly what I am talking about.

Room accommodations were in Sleeper Hall, which has a large lounge on the ground floor where everybody gathered for discussion of the events of the day and of the conference in general. Dining accommodations were in the adjacent building, Claflin Hall, and there more, animated discussions were conducted.

In that lounge I met Klaus Barner, who as a student attended many of Siegel's lectures in Germany. He expressed significant insight into the personality of this renowned mathematician, whom I had previously only known through his work. I spent much of the afternoon learning as much as I could about this fascinating man, who had retired somewhat early (1951) from his faculty position at the Institute for Advanced Study and returned to his native Germany where he continued to publish research papers (even after his Collected Works appeared) and to advise research students.

Not all conversation concentrated on mathematics, however. One well-known mathematician spent an entire lunch with Larry Washington and me telling us about his six months of jury duty in an important criminal trial. It was clear that he took his jury assignment very seriously and that he agonized over making as certain as possible that the final verdict was a just one.

All of the principals (except Serre and Taylor) involved in the proof were present for at least a portion of the conference as well as many of those in supporting roles, and without exception they all were very open, approachable and accommodating in their attitude. They bestowed considerable class to the gathering and set the tone for it.

It was all very, very exciting, and I only wish I had fluency in a language exceptionally rich in descriptive adjectives so that I could set down in words what I and many others had realized there. It was a singularly unique experience!

Not to neglect a charmingly commercial recognition of the wide interest in the Fermat developments, a T-shirt was created for the occasion and enjoyed a brisk sale. On its front it displayed a sketch of the proof, with a single, ingeniously accurate, crucial paragraph; on its back appeared a list of literature references to the fundamental papers by Frey, Serre, Ribet, Wiles and Wiles-Taylor. It is clear that these shirts will be worn with a sense of distinction for many years.

Also, several book publishers displayed their relevant books and journals at a large table in the lobby of the College. Ina Lindemann represented Springer-Verlag here and when Wiles spoke at Hunter College.

Rounding out the social aspects of the conference, a reception was held on August 12th at the George Sherman Union. The grand buffet and open bar were a fitting celebration of the spirit of the conference and of the cooperative work that had gone on for so many years, leading up to this event.

Late in the afternoon of August 17th my Leica M6, which had served me so faithfully during the past eighteen months, jammed; I had to finish the conference with my very noisy Nikon F3 backup reflex camera. I like to think that my determination to finish the job was another reflection of the dedication shown by the entire group to reach a common goal.

And through all this activity the Fermat work went on. Chris Skinner and Brian Conrad conducted a tutorial for those who desired to gain even more insight into the proof itself. Invited lectures were presented from Wednesday, August 9th through Friday, August 18th. All of the lectures were impeccably prepared and delivered. On Sunday, August 13th, a number of twenty-minute talks were presented in several rooms of the College of General Studies. Most of the invited speakers made available preprints of their conference papers, and all of the twenty-minute speakers provided detailed abstracts of their talks.

3. August 9, 1995

Wiles presented the last lecture of the conference.[2] At the end of that lecture (cf. Photograph 59) he thanked Ribet for providing him with seven years of work. He then said with regard to Frey, and with his typical wry humor, "If he has any ideas about this problem about zeroes and if he would like to share them with us, we would be happy to listen."

A question that was present at all times during the conference was, "How close are we to resolving the full Shimura-Taniyama conjecture?" The view of the experts was mixed, but the consensus seemed to be, "Not as close as we had initially thought... or hoped." One expert opined that it may take a fundamentally new idea to establish the full conjecture.

Until June 23, 1999 the best established result, due to the joint work of Conrad, Taylor and Diamond, was that if 27 does not divide the conductor, then the elliptic curve is modular (see [**17**]). However, on that date six years, almost to the hour, after Wiles had announced his claim at Cambridge University of the Shimura-Taniyama conjecture for semistable elliptic curves, I received the following e-mail from Larry Washington:

> I just got a message that said Ribet said that Conrad, Diamond, Taylor and Breuil have finished the proof of Taniyama. I don't have more details now. I'll let you know if I hear more.

Previously, on May 10, 1999, Diamond gave a talk at the Goldfeld-Jacquet-Zhang seminar at Columbia University. During tea I talked briefly with him about the present status of the Shimura-Taniyama conjecture. The next day, based on that conversation, I sent the following e-mail to several of my associates:

3. August 9, 1995

I talked with Diamond on Monday at Columbia, and I asked him how close are we to the full Shimura-Taniyama conjecture. He smiled and said, "Close." He then mentioned something about a new result from "somebody overseas" that will help. Let me emphasize that he made no claim of the full Shimura-Taniyama conjecture at that time.

According to Breuil, (cf. Photograph 60), to get their result they use many results of Raynaud on flat finite group schemes.[3]

The 1995–96 academic year at the Institute for Advanced Study was devoted to Wiles and his exceptionally difficult and very comprehensive proof, and more experts had the opportunity to lecture on various aspects of it. And Wiles continued to lecture in Europe and in America. But for many of us, who started on this long and arduous journey with him in Taplin Auditorium in early February of 1994 and shared some of the emotion, pressure and ultimate elation, it was over. We were saturated and exhausted, and it showed. But how very fortunate we were to have had the opportunity to witness this remarkable man in his gallant struggle to realize his childhood dream, a dream whose fulfillment has produced results that will have a profound influence on mathematics for many years to come.

42. Yves Hellegouarch at the Fermat Conference, College of General Studies Lobby, Boston University, August 1995.

3. August 9, 1995

43. Gerhard Frey, Fermat Conference, Jacob Sleeper Hall, Boston University, August 12, 1995.

44. Ken Ribet, Fermat Conference, Jacob Sleeper Hall, Boston University, August 1995.

45. Barry Mazur, Fermat Conference, Jacob Sleeper Hall, Boston University, August 1995.

3. August 9, 1995

46. John Tate, Fermat Conference, Jacob Sleeper Hall, Boston University, August 10, 1995.

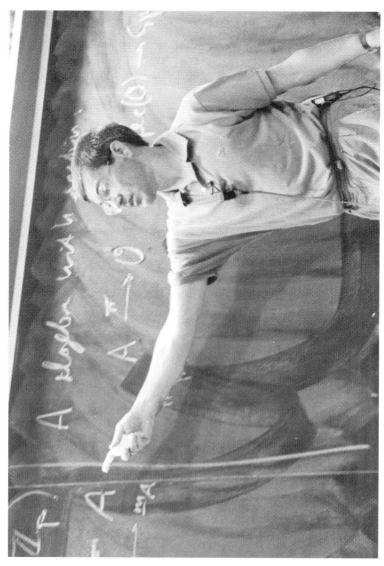

47. René Schoof, Fermat Conference, Jacob Sleeper Hall, Boston University, August 15, 1995.

3. August 9, 1995

48. Ehud de Shalit, Fermat Conference, Jacob Sleeper Hall, Boston University, August 17, 1995.

49. Jacques Tilouine, Fermat Conference, Jacob Sleeper Hall, Boston University, August 14, 1995.

3. August 9, 1995

50. Hendrik Lenstra, Fermat Conference, Jacob Sleeper Hall, Boston University, August 11, 1995.

51. Karl Rubin, Fermat Conference, Jacob Sleeper Hall, Boston University, August 18, 1995.

3. August 9, 1995

52. Benedict Gross, Fermat Conference, Jacob Sleeper Hall, Boston University, August 11, 1995.

53. Joe Silverman, Fermat Conference, Jacob Sleeper Hall, Boston University, August 10, 1995.

3. August 9, 1995

54. Glenn Stevens, Fermat Conference, Jacob Sleeper Hall, Boston University, August 9, 1995.

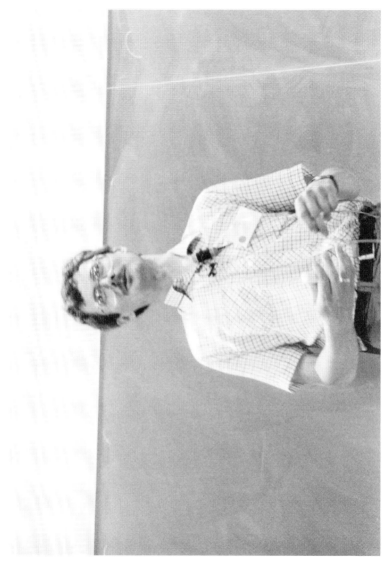

55. Jaap Top, Fermat Conference, Jacob Sleeper Hall, Boston University, August 10, 1995.

56. Alice Silverberg, Fermat Conference, Jacob Sleeper Hall, Boston University, August 17, 1995.

57. Michael Rosen, Fermat Conference, Jacob Sleeper Hall, Boston University, August 17, 1995.

3. August 9, 1995

58. Gary Cornell, Fermat Conference, Jacob Sleeper Hall, Boston University, August 15, 1998.

59. Andrew Wiles, Fermat Conference, Jacob Sleeper Hall, Boston University, August 18, 1995.

3. August 9, 1995

60. Christophe Breuil, Brive-la-Gaillarde, France, May 1998.

61. Dorian Goldfeld, Columbia University, February 10, 1997.

3. August 9, 1995

62. Ram Murty, Conference in Honor of Hans Rademacher, Penn State University, July 1992.

Cast of Characters

Positions indicated are at the time when the person is mentioned in the text.

Amir Aczel	Associate Professor of Statistics, Bentley College
James Arthur	Professor of Mathematics, University of Toronto
Abbas Bahri	Professor of Mathematics, Rutgers University
Klaus Barner	Professor of Mathematics, University of Kassel
Eric Temple Bell	(1883–1960) American Mathematician
Bryan Birch	Professor of Mathematics, Emeritus, University of Oxford
Morris Bobrow	Song Writer and Performer in San Francisco
Enrico Bombieri	Professor of Mathematics, Institute for Advanced Study
Lenore Blum	Deputy Director, Mathematical Sciences Research Institute

Cast of Characters

Christophe Breuil	C. N. R. S. Researcher, Université Paris-Sud
Justine Bumby	American Mathematician
Henri Cartier-Bresson	French Photographer
Sarva Daman Chowla	(1907–1995) Indian-American Mathematician
John Coates	Professor of Mathematics, Cambridge University
Brian Conrad	Graduate Student, Princeton University
John Conway	Professor of Mathematics, Princeton University
Gary Cornell	Associate Professor of Mathematics, The University of Connecticut
David Cox	Professor of Mathematics, Amherst College
Henri Darmon	Instructor of Mathematics, Princeton University
Pierre Deligne	Professor of Mathematics, Institute for Advanced Study
Lee Dembart	Editorial Page Editor, *San Francisco Examiner*
Fred Diamond	Ritt Assistant Professor of Mathematics, Columbia University
Diophantus of Alexandria	(AD 250) Greek Mathematician
Mary P. Dolciani	(1923–1985) American Mathematician
Edward Dunne	Editor for the Book Program, The American Mathematical Society
Gerd Faltings	Professor of Mathematics, Princeton University
Pierre de Fermat	(1601–1644) French Mathematician
John Charles Fields	(1863–1932) Canadian Mathematician
Matthias Flach	Visiting Mathematician, Princeton University

Cast of Characters

Jean-Marc Fontaine	Professor of Mathematics, Université de Paris-Sud
Gerhard Frey	Director, Institute for Experimental Mathematics, University of Essen
Évariste Galois	(1811–1832) French Mathematician
Carl Friedrich Gauss	(1777–1855) German Mathematician
Dorian Goldfeld	Professor of Mathematics, Columbia University
Fernando Gouvêa	Assistant Professor of Mathematics and Computer Science, Colby College
Ralph Greenberg	Professor of Mathematics, University of Washington
Phillip Griffiths	Director, Institute for Advanced Study
Benedict Gross	Professor of Mathematics, Harvard University
John Grothendieck	Undergraduate Student, Harvard University
Li Guo	Research Student, Ohio State University
Jacques Hadamard	(1865–1963) French Mathematician
G. H. Hardy	(1877–1947) British Mathematician
Will Hearst	Publisher and Editor of the *San Francisco Examiner*
William Randolph Hearst	(1863–1951) Newspaper Publisher
Yves Hellegouarch	Professor of Mathematics, Université de Caen
Luc Illusie	Professor of Mathematics, Université de Paris-Sud
Kenkichi Iwasawa	(1917–1998) Japanese-American Mathematician
Shokichi Iyanaga	Professor of Mathematics, Emeritus, Tokyo University
Allyn Jackson	Staff Writer, American Mathematical Society

Cast of Characters

Hervé Jacquet	Professor of Mathematics, Columbia University
Ian Katz	Science Writer, *The Guardian*
Nicholas Katz	Professor of Mathematics, Princeton University
Joseph J. Kohn	Professor of Mathematics, Princeton University
Gina Kolata	Science Writer, *The New York Times*
Victor Kolyvagin	Professor of Mathematics, The Johns Hopkins University
Ernst Eduard Kummer	(1810–1893) German Mathematician
Serge Lang	Professor of Mathematics, Yale University
Robert Langlands	Professor of Mathematics, Institute for Advanced Study
Tom Lehrer	Mathematician, Singer, Composer, Lyricist, and Entertainer
Hendrik Lenstra, Jr.	Professor of Mathematics, University of California, Berkeley
Ina Lindemann	Executive Editor, Mathematics, Springer-Verlag
Trevor Lipscombe	Physical Sciences Editor, Princeton University Press
John Lynch	Editor of BBC TV's Horizon Series
Robert MacPherson	Professor of Mathematics, Institute for Advanced Study
Attila Máté	Professor of Mathematics, Brooklyn College of the City University of New York
Barry Mazur	Professor of Mathematics, Harvard University
Jean-François Mestre	Professor of Mathematics, Université Paris 7
Kumar Murty	Professor of Mathematics, University of Toronto

Cast of Characters

Ram Murty	Professor of Mathematics, Queen's University
John Nash, Jr.	American Mathematician
Alfred Nobel	(1833–1896) Swedish Chemist, Engineer and Industrialist
Andrew Ogg	Professor of Mathematics, University of California, Berkeley
Robert Osserman	Deputy Director, Mathematical Sciences Research Institute
Ilya Piatetski-Shapiro	Professor of Mathematics, Yale University
Alf van der Poorten	Professor of Mathematics, Macquaire University, Australia
M. Raynaud	Professor of Mathematics, Université Paris-Sud
Kenneth Ribet	Professor of Mathematics, University of California, Berkeley
Georg Friedrich Bernhard Riemann	(1826–1866) German Mathematician
Joseph Roitberg	Professor of Mathematics, Hunter College
Michael Rosen	Professor of Mathematics, Brown University
Karl Rubin	Professor of Mathematics, Ohio State University
Zeev Rudnick	Assistant Professor of Mathematics, Princeton University
Peter Sarnak	Professor of Mathematics, Princeton University
René Schoof	Professor of Mathematics, Università di Roma, Italy
Jean-Pierre Serre	Professor of Mathematics, Collège de France

Joseph A. Shalika	Professor of Mathematics, The Johns Hopkins University
Ehud de Shalit	Institute of Mathematics, Hebrew University
Harold Shapiro	President, Princeton University
Goro Shimura	Professor of Mathematics, Princeton University
Carl Ludwig Siegel	(1896–1981) German-American Mathematician
Alice Silverberg	Associate Professor of Mathematics, Ohio State University
Joseph Silverman	Professor of Mathematics, Brown University
Yakov Sinai	Professor of Mathematics, Princeton University
Simon Singh	Science Department, BBC
Chris Skinner	Graduate Student, Princeton University
Glenn Stevens	Professor of Mathematics, Boston University
Peter Swinnerton-Dyer	Professor of Mathematics, Cambridge University
Lucien Szpiro	Directeur de Recherches (CNRS), École Normale Supérieure
Yutaka Taniyama	(1927–1958) Japanese Mathematician
John Tate	Professor of Mathematics, University of Texas at Austin
Richard Taylor	Professor of Mathematics, Cambridge University
Jacques Tilouine	Professor of Mathematics, Université de Paris 13
Jaap Top	Associate Professor of Mathematics, University of Groningen

Cast of Characters

Jerrold Tunnell	Associate Professor of Mathematics, Rutgers University
Nike Vatsal	Graduate Student, Princeton University
Lawrence Washington	Professor of Mathematics, University of Maryland
André Weil	Professor of Mathematics, Emeritus, Institute for Advanced Study
Hermann Weyl	(1885–1955) German-American Mathematician
Andrew Wiles	Professor of Mathematics, Princeton University
Paul Wolfskehl	(1856-1906) German Medical Doctor by Academic Degree and Mathematician by Avocation
Shing-Tung Yau	Professor of Mathematics, Harvard University
Shou-Wu Zhang	Assistant Professor of Mathematics, Princeton University

List of Photographs

1. Andrew Wiles 27
2. Andrew Wiles 28
3. Nicholas Katz 29
4. Barry Mazur 30
5. Jean-Marc Fontaine 31
6. Peter Sarnak 32
7. John Coates 33
8. Jean-Pierre Serre 34
9. Jean-Pierre Serre and John Tate 35
10. Gerd Faltings 36
11. Andrew Wiles and Gerd Faltings 37
12. Pierre Deligne with Phillip Griffiths 38
13. Henri Darmon 39
14. Fred Diamond 40
15. Larry Washington 41

List of Photographs

16. Shou-Wu Zhang 42
17. Lucien Szpiro 43
18. Chris Skinner and Nike Vatsal 44
19. Brian Conrad 45
20. Jerry Tunnell 46
21. Hervé Jacquet 47
22. Ilya Piatetski-Shapiro 48
23. Joseph A. Shalika 49
24. Robert Langlands 50
25. Andrew Wiles 51
26. Goro Shimura and André Weil 81
27. Andrew Wiles 82
28. Andrew Wiles 83
29. Andrew Wiles 84
30. Andrew Wiles 85
31. Richard Taylor 86
32. Andrew Wiles 87
33. Andrew Wiles 88
34. Andrew Wiles 89
35. Andrew Wiles 90
36. Andrew Wiles 91
37. Andrew Wiles 92
38. Andrew Wiles 93
39. Andrew Wiles 94

List of Photographs 135

40. Andrew Wiles 95
41. The Princeton home of Andrew Wiles 96
42. Yves Hellegouarch 103
43. Gerhard Frey 104
44. Ken Ribet 105
45. Barry Mazur 106
46. John Tate 107
47. René Schoof 108
48. Ehud de Shalit 109
49. Jacques Tilouine 110
50. Hendrik Lenstra 111
51. Karl Rubin 112
52. Benedict Gross 113
53. Joe Silverman 114
54. Glenn Stevens 115
55. Jaap Top 116
56. Alice Silverberg 117
57. Michael Rosen 118
58. Gary Cornell 119
59. Andrew Wiles 120
60. Christophe Breuil 121
61. Dorian Goldfeld 122
62. Ram Murty 123

Acknowledgements and Notes

Dorian Goldfeld convinced me to write this book, and he maintained a gentle but steady pressure on me over a period of four years until I ultimately did finish it. Simply said, if it were not for his efforts, the book would never have been written. I thank him, Peter Sarnak, Henryk Iwaniec and Larry Washington for their encouragement, suggestions and assistance.

Many people helped me with background material and details, and I thank them all, but special thanks go to Karl Rubin, Ken Ribet, Dorian Goldfeld, Shou-Wu Zhang, Hervé Jacquet, Larry Washington, Emmanuel Kowalski and Li Guo, who also kindly supplied Photograph 1.

I am indebted to Klaus Barner for the cover photograph and to Christophe Breuil for Photograph 60.

Ram Murty's permission to reproduce his review is most appreciated.

Joseph Kohn was the chairman of the Princeton Mathematics Department during 1994–95, and I thank him for his kindness and forbearance and for allowing me to use the facilities of the department.

Acknowledgements and Notes

The e-mails included in the book from Gerhard Frey, Ken Ribet, Gerd Faltings, Dorian Goldfeld, Larry Washington and Christophe Breuil are most appreciated.

Conversations with Edward Dunne, Attila Máté and Trevor Lipscombe concerning style and form were most helpful.

I thank Andrew Wiles for allowing me to attend his lectures and those of his graduate students, for permitting me to reproduce a portion of his Annals paper in Appendix A, and for his cordiality during every encounter that I have had with him.

The manuscript was typeset in $\mathcal{A}_{\mathcal{M}}\mathcal{S}$-TEX by Dottie Phares of the Institute for Advanced Study, and the photographs were printed by Donna Gruber. Their splendid craftmanship was crucial for the successful production of the book.

Finally, I am indebted to Steve Miller for painstakingly proofreading the manuscript.

Note for Preface

1. There is some disagreement as to which of the mathematicians, Shimura, Taniyama, Weil, should be associated with the conjecture.

In the November 1995 issue of the *Notices of the American Mathematical Society*, Serge Lang published a lengthy and characteristically carefully-documented article entitled, "Some History of the Shimura-Taniyama Conjecture." Professor Lang informed me that his purpose for writing the article was to bring some bibliographical items to the attention of the mathematics community. This article and the articles mentioned in its representative bibliography would be a good starting point for those who are interested in pursuing this matter further.

At this time most experts only associate Shimura and Taniyama with the conjecture.

On February 11, 2000 Shimura (cf. Photograph 26) explained to me his position on the matter:

In 1955 Yutaka Taniyama [who committed suicide in 1958 at the age of 31] raised questions in which he indicated the possibility that all elliptic curves are uniformized by a class of automorphic functions wider than modular functions.

Around 1964 Shimura conjectured that every elliptic curve defined over the rational number field is uniformized by modular functions.

Shimura never collaborated with Taniyama on the conjecture although he did collaborate with him on other topics.

In the April 1999 issue of the *Notices of the American Mathematical Society*, Shokichi Iyanaga includes in his essay a remark on the origin of the conjecture.

References [1], [115], [116] and [117] contain information relevant to this matter, also.

A popular exposition of the conjecture, understandable to the layperson, together with a brief history of the conjecture itself, appears in [40].

Notes for Chapter 1

1. The Institute for Advanced Study in Princeton, NJ was founded in 1930 by a gift from Mr. Louis Bamberger and his sister, Mrs. Felix Fuld. It is located just west of the campus of Princeton University. Albert Einstein was a faculty member from 1933 until his death in 1955. The faculty has included, among others, Hermann Weyl, Kurt Gödel, John von Neumann, Marston Morse, Atle Selberg, Harish-Chandra, Carl Ludwig Siegel, Deane Montgomery, Armand Borel and André Weil. Presently there are four separate schools at the Institute: The School of Historical Studies, The School of Mathematics, The School of Natural Sciences, and The School of Social Sciences.

2. By careful and elegant exposition, Ken Ribet has made a concentrated effort to make clear his work and some aspects of Wiles's work needed to establish Fermat's Last Theorem. In particular, the reader is referred to [26], [90], [93], [94], [96], [97], [98], and [117].

3. The Mathematical Sciences Research Institute (MSRI) is located adjacent to the campus of the University of California at Berkeley.

4. Enrico Bombieri, who has been awarded the Fields Medal, has made substantial contributions to differential geometry, analytic geometry, algebraic geometry, number theory and finite groups. Further, he is an accomplished artist who exhibits and sells his work. He once told me that he pursues his art with the same intensity that he pursues his mathematics. Cf. Note 25, Chapter 1 for a discussion of the Fields Medal.

5. Cf. Cox [20], Gouvêa [41] and Rubin-Silverberg [103].

Copies of the Ribet and the Mazur videos were immediately circulated throughout the mathematics community. They were subsequently published by the American Mathematical Society (cf. [73], [94]).

The Gouvêa paper went on to win the Lester R. Ford Award as one of the best articles published in the *Monthly* that year.

The Lester R. Ford Awards were established in 1964 to recognize authors of articles of expository excellence published in *The American Mathematical Monthly* or *Mathematics Magazine*. Beginning in 1976, a separate award (the Allendoerfer Award) was created for *Mathematics Magazine*. The awards are named for Lester R. Ford, Sr., a distinguished mathematician, editor of *The American Mathematical Monthly*, 1942–1946, and President of the Mathematical Association of America, 1947–1948. This is an award of $500. Up to five of these awards are given annually at the Summer Meeting of the Association. All awards given from 1976 on are for articles that appeared in *The American Mathematical Monthly*. Between 1965 and 1975,

Notes for Chapter 1

Ford awards were given for articles in the *Monthly* or *Mathematics Magazine*.

6. For the most accurate account available of the scenario the reader is referred to the Introduction in [130]. We have included an excerpt from it as Appendix A. Mastery of the lead article in [19] (G. Stevens, *An overview of the proof of Fermat's Last Theorem*) and of the lead article in [15] (H. Darmon, F. Diamond and R. Taylor, *Fermat's Last Theorem*) would provide orientation background to start a serious reading of [119] and [130] themselves.

7. The literature on Fermat's Last Theorem is enormous. References [34], [87] and [88] are highly recommended. On several occasions I heard Wiles state that in his youth, he felt that Fermat could not have known very much more about the problem than he did.

8. On November 27, 1944 the University of Freiburg in Germany was very badly damaged by a wartime bombing raid, and the mathematics department was subsequently relocated to a hunting lodge in Oberwolfach, a small rural town northeast of Freiburg. This lodge evolved into what is now known as the Mathematisches Forschungsinstitut Oberwolfach (English translation: Mathematical Research Institute Oberwolfach). I was invited to speak there in April 1984 (and again in March 1991), and I found that the hunting lodge had been replaced by a large complex of modern and efficient buildings. The Institute publishes in German (and in English translation) a very interesting 56-page monograph devoted to the history of the Institute. On the cover appears a photograph of the original hunting lodge. In the August 2000 issue of the *Notices of the American Mathematical Society* Allyn Jackson published an interesting article entitled, "Oberwolfach, Yesterday and Today."

9. E-mail received from Ken Ribet dated June 30, 1998.

10. Cf. Aczel [1], Singh [116] and van der Poorten [83].

11. E-mail received from Gerhard Frey dated August 27, 1998. Dorian Goldfeld (cf. Photograph 61), who in 1987 won the Cole Prize in number theory (jointly with Benedict Gross and Don Zagier) for solving Gauss's class-number problem, has presented a careful discussion of the ABC conjecture in [40]. Cf. Note 13, Chapter 2 for a discussion of the Cole Prize.

12. BBC *Horizon* broadcast of January 15, 1996. John Lynch and Simon Singh conducted a series of lengthy, in depth, video-recorded interviews of most of the principals involved in the proof of Fermat's Last Theorem. Then they skillfully edited this mass of material into a delightful and informative BBC broadcast (this program was reviewed in the January 1997 issue of the *Notices of the American Mathematical Society*) that aired in the *Horizon* series on January 15, 1996 entitled "Fermat's Last Theorem" (this was rebroadcast in America on NOVA several times starting on October 28, 1997 under the title "The Proof"). Some of the interview material was also incorporated in Singh's book [116] (this book was reviewed in the November 1997 issue of the *Notices of the American Mathematical Society*).

In 1999 Lynch and Singh jointly received The Special Communications Award from the Joint Policy Board of Mathematics (JPBM) consisting of representative members of the American Mathematical Society (AMA), the Mathematical Association of America (MAA) and the Society of Industrial and Applied Mathematics (SIAM). The award recognizes individuals who bring accurate mathematical information to nonmathematicians, a prime example being its award in 1996 to Gina Kolata, well known for her writing to large audiences in her *New York Times* articles.

13. Serre's letter was published in the proceedings of the AMS-IMS-SIAM Joint Summer Research Conference, which was held August 18–24, 1985 at Humboldt State University in Arcata, California. Ken Ribet was the editor of that volume which appears as volume 67 in the American Mathematical Society's "Contemporary Mathematics"

Notes for Chapter 1   143

series. In Appendix D there is a brief technical discussion of Serre's conjectures.

14. Ribet's paper appeared as [91]; see also [90].

15. The Prix Fermat was created by Paul Sabatier University (University of Toulouse III, France) to be under the patronage of the Matra Marconi Space Company. The amount of the prize has been fixed at FF100,000. The prize is awarded for research in fields where the contributions of Pierre Fermat have been decisive.

16. BBC *Horizon* broadcast of January 15, 1996.

17. In the spring of 1985 Dorian Goldfeld met Zhang at the Chinese Academy of Sciences in Beijing. At the time he was a student there from a small village near Nanjing. Goldfeld subsequently invited him to enroll in the Columbia University Graduate School; Zhang did so in the fall in 1986 and received his Ph.D. in June of 1991. He wrote his thesis under the direction of Lucien Szpiro and Gerd Faltings. He spent four years on the faculty of Princeton Unviersity and one year as a member at the Institute for Advanced Study. In 1998 he was awarded the Morningside Gold Medal. This medal was created with the support of the Chinese Academy of Sciences and a generous donation from the Morningside Group.

18. BBC *Horizon* broadcast of January 15, 1996.

19. BBC *Horizon* broadcast of January 15, 1996.

20. Appendix B contains a list of the speakers at the conference.

21. Interview with Li Guo April 23, 1999. The statement that Wiles wrote on the board is equivalent to Fermat's Last Theorem as stated in the Preface. To prove that equivalence one needs the fact that Fermat's Last Theorem is true for $n = 4$, which can be established by a very elementary argument (cf. Section 1.5 in [34]).

22. In an e-mail dated October 23, 1999, Dorian Goldfeld explained it this way:

Weil's paper "Jacobi Sums as Grossencharaktere" *Trans. Amer. Math. Soc.* VI. **73**, 487–495 (1952) was the first paper to prove that for certain CM elliptic curves over **Q** the Hasse-Weil L-function is a Hecke L-function with Grossencharakter. This result was generalized to all CM elliptic curves by Deuring, *Die Zetafunktion einer algebraischen Kurve von Geschlechte Eins*, Nachr. Akad. Wiss. Göttingen, Math-Physik (1953).

In his 1967 paper "Uber die Bestimmung Dirichletscher Reihen durch Funktionalgleichungen" *Math. Ann.* 168 (1967) Weil proves the first converse theorem which combined with Deuring's paper proves that all CM curves defined over **Q** are modular.

It can be shown that each such CM curve can be associated with one and only one of nine j-invariants (Cf. A. Borel, S. Chowla, C. S. Herz, K. Iwasawa, J.-P. Serre, *Seminar on Complex Multiplication*, LNM #21, Springer-Verlag, 1966).

23. Cf. Langlands [66], Tunnell [122] and Jacquet et al. [57].

Robert Langlands has shaped the modern theory of automorphic forms, and he has proposed a significant unification of mathematics popularly referred to as "the Langlands program." For a nice introduction to and orientation in this program the reader is referred to [39]. He has also made substantial contributions to noncommutative analysis and number theory. Much of his important work is available at the web site (http://sunsite.ubc.ca). Cf. Note 5 Chapter 2 for a discussion of the Steele Prize.

24. BBC *Horizon* broadcast of January 15, 1996.

25. At the 1924 International Congress of Mathematicians held in Toronto a resolution was adopted whereby at each Congress gold

medals should be awarded to recognize outstanding mathematical achievement by young mathematicians, where "young" has been traditionally interpreted to mean no more than forty years of age in the year of the Congress, which is held every four years at various locations throughout the world. J. C. Fields, a Canadian mathematician, who was secretary of the 1924 Congress, donated funds establishing the medals, which were named in his honor. The Fields Medal is generally considered to be the world's highest honor in mathematics.

There are various theories in the folklore as to why under the terms of the will of Alfred Nobel there is no Nobel Prize awarded in mathematics or in astronomy. The two most popular ones are (1) Nobel was a very "practical" man who saw little use for astronomy and mathematics, especially pure mathematics. (2) Nobel had had a very serious personal dispute with the Swedish mathematician, and astronomer, Gosta Mittag-Leffler. I spent the 1977–78 academic year in residence at the Mittag-Leffler Institute in Djursholm, Sweden, and I tried without success to determine the truth of the matter. To further complicate matters, Mittag-Leffler actually nominated people for the Nobel prize in physics.

Notes for Chapter 2

1. BBC *Horizon* broadcast of January 15, 1996.

2. E-mail received from Gerd Faltings dated February 14, 1996.

3. For a nice introduction to and orientation in the deep and important work of Faltings, for which he received a Fields Medal in 1986, the reader is referred to [18], [64], [108], and [118]. Cf. Note 25, Chapter 1 for a discussion of the Fields Medal.

4. BBC *Horizon* broadcast of January 15, 1996.

5. Nash received the Nobel prize for his work in game theory, which was considered to be in the field of economics rather than mathematics. Nash's early and brilliant career in mathematics was tragically

brought to a standstill by a terrible illness. He is recovered now, and some of his fundamental research is finally being properly recognized. In 1999 he received the Leroy P. Steele prize for a seminal contribution to research. The Steele prize was established in 1970 in honor of George David Birkhoff, William Fogg Osgood, and William Gasper Graustein and is endowed under the terms of a bequest from Leroy P. Steele. The current award is $4000. Recently, an unauthorized biography of Nash has appeared, which contains a lot of irrelevant, unsubstantiated gossip. For a serious discussion of his scientific work the reader is referred to [104].

6. In his introduction to Wiles's lectures, Professor Roitberg stated:

> Professor Dolciani was a longtime member of the Hunter College faculty. In 1985 she and her husband, James Halloran, established the Dolciani Halloran Foundation. Subsequently, the foundation has been involved in various activities at the college. In particular, the department benefited from very generous grants from the foundation that have made possible the setting up of a modern mathematics learning center/computer lab and also this lecture series.
>
> As for Mary Dolciani's professional career, it is certainly true that she is best known as a Mathematics educator, textbook writer and administrator. What is perhaps not as well-known is that before making the career decision to concentrate on educational matters, she had begun a career as a researcher in number theory. After taking her doctoral degree at Cornell, she held postdoctoral positions at Oxford and the Institute for Advanced Study, where she immersed herself in studying the then-current methods of Hardy and Littlewood, Vinogradov, C. L. Siegel and others. She obtained generalizations

Notes for Chapter 2

of known results about quadratic forms in general, ternary quadratic forms in particular and Waring's problem.

7. The literature on the Riemann hypothesis, the generalized Riemann hypothesis and the problem concerning the existence or nonexistence of exceptional zeros is vast, but I would be remiss if I did not call the reader's attention to [33] for those seeking a basic, leisurely exposition and to [54], [55], [60], and [61] for the experts.

P. Deligne in [21], [22] proved the Riemann hypothesis for varieties over finite fields, a very significant achievement for which he received a Fields Medal in 1978. Cf. Note 25, Chapter 1 for a discussion of the Fields Medal.

8. BBC *Horizon* broadcast of January 15, 1996.

9. This is an exact quotation (cf. page 252 in [82]) of the notorious Latin statement that Fermat wrote in the margin of his copy of Book II of the *Arithmetica*, written (c. AD 250) by Diophantus of Alexandria. The following is an English translation of the statement:

> It is impossible for a cube to be written as a sum of two cubes or a fourth power to be written as the sum of two fourth powers or, in general, for any number which is a power greater than the second to be written as a sum of two like powers; I have a truly marvelous demonstration of this proposition which this margin is too narrow to contain.

10. BBC *Horizon* broadcast of January 15, 1996.

11. The Schock Prize was awarded by the Royal Swedish Academy of Sciences. Cf. Note 15, Chapter 1 for information about the Prix Fermat.

12. The Wolf Prize carries a monetary award of $100,000. Langlands received the Wolf Prize "for his path-blazing work and extraordinary insights in the fields of number theory, automorphic forms and group representation." Wiles received the Wolf Prize "for spectacular contributions to number theory and related fields, for major advances on fundamental conjectures, and for settling Fermat's Last Theorem." The Wolf Foundation was founded by the late Ricardo Wolf. The purpose of the foundation is to "promote science and art for the benefit of mankind."

In the National Academy of Sciences Award in Mathematics, Wiles was cited for "his proof of Fermat's Last Theorem, by discovering a beautiful strategy to establish a major portion of the Shimura-Taniyama conjecture, and for his courage and technical power in bringing his idea to completion."

13. The Frank Nelson Cole Prize in Number Theory was founded in honor of Professor Frank Cole on the occasion of his retirement as secretary of the American Mathematical Society after twenty-five years and as editor-in-chief of the *Bulletin* for twenty-one years. The prize was awarded to Wiles "for his work on the Shimura-Taniyama conjecture and Fermat's Last Theorem."

In January 1905 Paul Wolfskehl, who had attended lectures in Berlin on the Fermat problem by Ernst Eduard Kummer, himself, altered his last will and testament in favor of "whomever first succeeds in proving the Great Theorem of Fermat." In present German currency he left DM75,000, and he directed that the Royal Society of Science in Göttingen should hold the money in trust and serve as judge for the awarding of the prize.

In the November 1997 issue of the *Notices of the American Mathematical Society* Klaus Barner published an interesting article entitled, "Paul Wolfskehl and the Wolfskehl Prize."

The MacArthur fellowship was from the John D. and Catherine T. MacArthur Foundation, and Wiles received a stipend totaling

$275,000 to be distributed over five consecutive years. It was awarded to Wiles for his work on the Shimura-Taniyama conjecture and for his work on Fermat's Last Theorem.

14. For his proof of Fermat's Last Theorem, Wiles received the King Faisal International Prize for Science. The prize consists of a $200,000 cash award and a commemorative gold medal, and it was awarded to Wiles by the King Faisal Foundation in a special ceremony in Riyadh, Saudi Arabia.

Wiles was too old to qualify for a Fields Medal so the International Mathematical Union awarded to him a one-time "Special Tribute" in the form of an "IMU Silver Plaque" during the opening ceremonies of the 1998 Congress in Berlin in recognition of his work that led to the proof of Fermat's Last Theorem. Wiles then received a prolonged standing ovation from the participants at the Congress that lasted several minutes. Cf. Note 25, Chapter 1 for a discussion of the Fields Medal.

Notes for Chapter 3

1. See also the nice review by Kevin Buzzard in Bull. Amer. Math. Soc. (N.S.) **36** (1999), 261–266.

2. Appendix C contains a list of the speakers at the conference.

3. E-mail received from Christophe Breuil dated July 19, 1999.

In the December 1999 issue of the *Notices of the American Mathematical Society* Henri Darmon published an interesting article entitled, "A Proof of the Full Shimura-Taniyama-Weil Conjecture Is Announced."

Breuil was a Ph.D. student of Jean-Marc Fontaine (cf. Photograph 5), and Diamond was a Ph.D. student of Andrew Wiles.

Appendix A

An Excerpt From the Introduction to Wiles's Annals of Mathematics Paper

Introduction

An elliptic curve over $\mathbf{Q}$ is said to be modular if it has a finite covering by a modular curve of the form $X_0(N)$. Any such elliptic curve has the property that its Hasse-Weil zeta function has an analytic continuation and satisfies a functional equation of the standard type. If an elliptic curve over $\mathbf{Q}$ with a given j-invariant is modular then it is easy to see that all elliptic curves with the same j-invariant are modular (in which case we say that the j-invariant is modular). A well-known conjecture which grew out of the work of Shimura and Taniyama in the 1950's and 1960's asserts that every elliptic curve over **Q** is modular. However, it only became widely known through its publication in a paper of Weil in 1967 [125] (as an exercise for the interested reader!), in which, moreover, Weil gave conceptual evidence for the conjecture. Although it had been numerically verified in many cases, prior to the results described in this paper it had only been known that finitely many j-invariants were modular.

In 1985 Frey made the remarkable observation that this conjecture should imply Fermat's Last Theorem. The precise mechanism relating the two was formulated by Serre as the ϵ-conjecture and this was then proved by Ribet in the summer of 1986. Ribet's result only requires one to prove the conjecture for semistable elliptic curves in order to deduce Fermat's Last Theorem.

Our approach to the study of elliptic curves is via their associated Galois representations. Suppose that ρ_p is the representation of $\text{Gal}(\overline{\mathbf{Q}}/\mathbf{Q})$ on the p-division points of an elliptic curve over **Q**, and suppose for the moment that ρ_3 is irreducible. The choice of 3 is critical because a crucial theorem of Langlands and Tunnell shows that if ρ_3 is irreducible then it is also modular. We then proceed by showing that under the hypothesis that ρ_3 is semistable at 3, together with some milder restrictions on the ramification of ρ_3 at the other primes, every suitable lifting of ρ_3 is modular. To do this we link the problem, via some novel arguments from commutative algebra, to a class number problem of a well-known type. This we then solve with the help of the paper [119]. This suffices to prove the modularity of E as it is known that E is modular if and only if the associated 3-adic representation is modular.

The key development in the proof is a new and surprising link between two strong but distinct traditions in number theory, the relationship between Galois representations and modular forms on the one hand and the interpretation of special values of L-functions on the other. The former tradition is of course more recent. Following the original results of Eichler and Shimura in the 1950's and 1960's

Introduction

the other main theorems were proved by Deligne, Serre and Langlands in the period up to 1980. This included the construction of Galois representations associated to modular forms, the refinements of Langlands and Deligne (later completed by Carayol), and the crucial application by Langlands of base change methods to give converse results in weight one. However with the exception of the rather special weight one case, including the extension by Tunnell of Langlands' original theorem, there was no progress in the direction of associating modular forms to Galois representations. From the mid 1980's the main impetus to the field was given by the conjectures of Serre which elaborated on the ϵ-conjecture alluded to before. Besides the work of Ribet and others on this problem we draw on some of the more specialized developments of the 1980's, notably those of Hida and Mazur.

The second tradition goes back to the famous analytic class number formula of Dirichlet, but owes its modern revival to the conjecture of Birch and Swinnerton-Dyer. In practice however, it is the ideas of Iwasawa in this field on which we attempt to draw, and which to a large extent we have to replace. The principles of Galois cohomology, and in particular the fundamental theorems of Poitou and Tate, also play an important role here.

The restriction that ρ_3 be irreducible at 3 is bypassed by means of an intriguing argument with families of elliptic curves which share a common ρ_5. Using this, we complete the proof that all semistable elliptic curves are modular. In particular, this finally yields a proof of Fermat's Last Theorem. In addition, this method seems well suited to establishing that all elliptic curves over $\mathbf{Q}$ are modular and to generalization to other totally real number fields.

...

The following is an account of the origins of this work and of the more specialized developments of the 1980's that affected it. I began working on these problems in the late summer of 1986 immediately on learning of Ribet's result. For several years I had been working on the Iwasawa conjecture for totally real fields and some applications of it. In the process, I had been using and developing results on ℓ-adic representations associated to Hilbert modular forms. It was therefore natural for me to consider the problem of modularity from the point of view of ℓ-adic representations. I began with the assumption that the reduction of a given ordinary ℓ-adic representation was reducible and tried to prove under this hypothesis that the representation itself would have to be modular. I hoped rather naively that in this situation I could apply the techniques of Iwasawa theory. Even more optimistically I hoped that the case $\ell = 2$ would be tractable as this would suffice for the study of the curves used by Frey. From now on and in the main text, we write p for ℓ because of the connections with Iwasawa theory.

After several months studying the 2-adic representation, I made the first real breakthrough in realizing that I could use the 3-adic representation instead: the Langlands-Tunnell theorem meant that ρ_3, the mod 3 representation of any given elliptic curve over $\mathbf{Q}$, would necessarily be modular. This enabled me to try inductively to prove that the $\mathrm{GL}_2(\mathbf{Z}/3^n\mathbf{Z})$ representation would be modular for each n. At this time I considered only the ordinary case. This led quickly to the study of $H^i(\mathrm{Gal}(F_\infty/\mathbf{Q}), W_f)$ for $i = 1$ and 2, where F_∞ is the splitting field of the $\mathfrak{m}$-adic torsion on the Jacobian of a suitable modular curve, $\mathfrak{m}$ being the maximal ideal of a Hecke ring associated to ρ_3 and W_f the module associated to a modular form f described in Chapter 1. More specifically, I needed to compare this cohomology with the cohomology of $\mathrm{Gal}(\mathbf{Q}_\Sigma/\mathbf{Q})$ acting on the same module.

I tried to apply some ideas from Iwasawa theory to this problem. In my solution to the Iwasawa conjecture for totally real fields [129], I had introduced a new technique in order to deal with the trivial zeroes. It involved replacing the standard Iwasawa theory method of considering the fields in the cyclotomic $\mathbf{Z}_p$-extension by a similar analysis based on a choice of infinitely many distinct primes $q_i \equiv 1$ mod p^{n_i} with $n_i \to \infty$ as $i \to \infty$. Some aspects of this method

Introduction

suggested that an alternative to the standard technique of Iwasawa theory, which seemed problematic in the study of W_f, might be to make a comparison between the cohomology groups as Σ varies but with the field $\mathbf{Q}$ fixed. The new principle said roughly that the unramified cohomology classes are trapped by the tamely ramified ones. After reading the paper [43], I realized that the duality theorems in Galois cohomology of Poitou and Tate would be useful for this. The crucial extract from this latter theory is in Section 2 of Chapter 1.

In order to put these ideas into practice I developed in a naive form the techniques of the first two sections of Chapter 2. This drew in particular on a detailed study of all the congruences between f and other modular forms of differing levels, a theory that had been initiated by Hida and Ribet. The outcome was that I could estimate the first cohomology group well under two assumptions, first that a certain subgroup of the second cohomology group vanished and second that the form f was chosen at the minimal level for $\mathfrak{m}$. These assumptions were much too restrictive to be really effective but at least they pointed in the right direction. Some of these arguments are to be found in the second section of Chapter 1 and some form the first weak approximation to the argument in Chapter 3. At that time, however, I used auxiliary primes $q \equiv -1 \mod p$ when varying Σ as the geometric techniques I worked with did not apply in general for primes $q \equiv 1 \mod p$. (This was for much the same reason that the reduction of level argument in [91] is much more difficult when $q \equiv 1 \mod p$.) In all this work I used the more general assumption that ρ_p was modular rather than the assumption that $p = 3$.

In the late 1980's, I translated these ideas into ring-theoretic language. A few years previously Hida had constructed some explicit one-parameter families of Galois representations. In an attempt to understand this, Mazur had been developing the language of deformations of Galois representations. Moreover, Mazur realized that the universal deformation rings he found should be given by Hecke rings, at least in certain special cases. This critical conjecture refined the expectation that all ordinary liftings of modular representations should be modular. In making the translation to this ring-theoretic language I realized that the vanishing assumption on the subgroup of H^2 which I had needed should be replaced by the stronger condition that the Hecke rings were complete intersections. This fitted well with their being deformation rings where one could estimate the number of generators and relations and so made the original assumption more plausible.

To be of use, the deformation theory required some development. Apart from some special examples examined by Boston and Mazur there had been little work on it. I checked that one could make the appropriate adjustments to the theory in order to describe deformation theories at the minimal level. In the fall of 1989, I set Ramakrishna, then a student of mine at Princeton, the task of proving the existence of a deformation theory associated to representations arising from finite flat group schemes over $\mathbf{Z}_p$. This was needed in order to remove the restriction to the ordinary case. These developments are described in the first section of Chapter 1 although the work of Ramakrishna was not completed until the fall of 1991. For a long time the ring-theoretic version of the problem, although more natural, did not look any simpler. The usual methods of Iwasawa theory when translated into the ring-theoretic language seemed to require unknown principles of base change. One needed to know the exact relations between the Hecke rings for different fields in the cyclotomic $\mathbf{Z}_p$-extension of $\mathbf{Q}$, and not just the relations up to torsion.

The turning point in this and indeed in the whole proof came in the spring of 1991. In searching for a clue from commutative algebra I had been particularly struck some years earlier by a paper of Kunz [62]. I had already needed to verify that the Hecke rings were Gorenstein in order to compute the congruences developed in Chapter 2. This property had first been proved by Mazur in the case of prime level and his argument had already been extended by other

Introduction

authors as the need arose. Kunz's paper suggested the use of an invariant (the η-invariant of the appendix) which I saw could be used to test for isomorphisms between Gorenstein rings. A different invariant (the $\mathfrak{p}/\mathfrak{p}^2$-invariant of the appendix) I had already observed could be used to test for isomorphisms between complete intersections. It was only on reading Section 6 of [121] that I learned that it followed from Tate's account of Grothendieck duality theory for complete intersections that these two invariants were equal for such rings. Not long afterwards I realized that, unlikely though it seemed at first, the equality of these invariants was actually a criterion for a Gorenstein ring to be a complete intersection. These arguments are given in the appendix.

The impact of this result on the main problem was enormous. Firstly, the relationship between the Hecke rings and the deformation rings could be tested just using these two invariants. In particular I could provide the inductive argument of Section 3 of Chapter 2 to show that if all liftings with restricted ramification are modular then all liftings are modular. This I had been trying to do for a long time but without success until the breakthrough in commutative algebra. Secondly, by means of a calculation of Hida summarized in [51] the main problem could be transformed into a problem about class numbers of a type well-known in Iwasawa theory. In particular, I could check this in the ordinary CM case using the recent theorems of Rubin and Kolyvagin. This is the content of Chapter 4. Thirdly, it meant that for the first time it could be verified that infinitely many j-invariants were modular. Finally, it meant that I could focus on the minimal level where the estimates given by my earlier Galois cohomology calculations looked more promising. Here I was also using the work of Ribet and others on Serre's conjecture (the same work of Ribet that had linked Fermat's Last Theorem to modular forms in the first place) to know that there was a minimal level.

The class number problem was of a type well-known in Iwasawa theory and in the ordinary case had already been conjectured by

Coates and Schmidt. However, the traditional methods of Iwasawa theory did not seem quite sufficient in this case and, as explained earlier, when translated into the ring-theoretic language seemed to require unknown principles of base change. So instead I developed further the idea of using auxiliary primes to replace the change of field that is used in Iwasawa theory. The Galois cohomology estimates described in Chapter 3 were now much stronger, although at that time I was still using primes $q \equiv -1 \mod p$ for the argument. The main difficulty was that although I knew how the η-invariant changed as one passed to an auxiliary level from the results of Chapter 2, I did not know how to estimate the change in the $\mathfrak{p}/\mathfrak{p}^2$-invariant precisely. However, the method did give the right bound for the generalised class group, or Selmer group as it is often called in this context, under the additional assumption that the minimal Hecke ring was a complete intersection.

I had earlier realized that ideally what I needed in this method of auxiliary primes was a replacement for the power series ring construction one obtains in the more natural approach based on Iwasawa theory. In this more usual setting, the projective limit of the Hecke rings for the varying fields in a cyclotomic tower would be expected to be a power series ring, at least if one assumed the vanishing of the μ-invariant. However, in the setting with auxiliary primes where one would change the level but not the field, the natural limiting process did not appear to be helpful, with the exception of the closely related and very important construction of Hida [49]. This method of Hida often gave one step towards a power series ring in the ordinary case. There were also tenuous hints of a patching argument in Iwasawa theory ([106], [129, §10]), but I searched without success for the key.

Then, in August, 1991, I learned of a new construction of Flach [35] and quickly became convinced that an extension of his method was more plausible. Flach's approach seemed to be the first step towards the construction of an Euler system, an approach which would give the precise upper bound for the size of the Selmer group if it could

Introduction

be completed. By the fall of 1992, I believed I had achieved this and began then to consider the remaining case where the mod 3 representation was assumed reducible. For several months I tried simply to repeat the methods using deformation rings and Hecke rings. Then unexpectedly in May 1993, on reading of a construction of twisted forms of modular curves in a paper of Mazur [71], I made a crucial and surprising breakthrough: I found the argument using families of elliptic curves with a common ρ_5 which is given in Chapter 5. Believing now that the proof was complete, I sketched the whole theory in three lectures in Cambridge, England on June 21–23. However, it became clear to me in the fall of 1993 that the construction of the Euler system used to extend Flach's method was incomplete and possibly flawed.

Chapter 3 follows the original approach I had taken to the problem of bounding the Selmer group but had abandoned on learning of Flach's paper. Darmon encouraged me in February, 1994, to explain the reduction to the complete intersection property, as it gave a quick way to exhibit infinite families of modular j-invariants. In presenting it in a lecture at Princeton, I made, almost unconsciously, a critical switch to the special primes used in Chapter 3 as auxiliary primes. I had only observed the existence and importance of these primes in the fall of 1992 while trying to extend Flach's work. Previously, I had only used primes $q \equiv -1 \mod p$ as auxiliary primes. In hindsight this change was crucial because of a development due to de Shalit. As explained before, I had realized earlier that Hida's theory often provided one step towards a power series ring at least in the ordinary case. At the Cambridge conference de Shalit had explained to me that for primes $q \equiv 1 \mod p$ he had obtained a version of Hida's results. But except for explaining the complete intersection argument in the lecture at Princeton, I still did not give any thought to my initial approach, which I had put aside since the summer of 1991, since I continued to believe that the Euler system approach was the correct one.

Meanwhile in January, 1994, R. Taylor had joined me in the attempt to repair the Euler system argument. Then in the spring of 1994, frustrated in the efforts to repair the Euler system argument, I began to work with Taylor on an attempt to devise a new argument using $p = 2$. The attempt to use $p = 2$ reached an impasse at the end of August. As Taylor was still not convinced that the Euler system argument was irreparable, I decided in September to take one last look at my attempt to generalize Flach, if only to formulate more precisely the obstruction. In doing this I came suddenly to a marvelous revelation: I saw in a flash on September 19th, 1994, that de Shalit's theory, if generalised, could be used together with duality to glue the Hecke rings at suitable auxiliary levels into a power series ring. I had unexpectedly found the missing key to my old abandoned approach. It was the old idea of picking q_i's with $q_i \equiv 1 \mod p^{n_i}$ and $n_i \to \infty$ as $i \to \infty$ that I used to achieve the limiting process. The switch to the special primes of Chapter 3 had made all this possible.

After I communicated the argument to Taylor, we spent the next few days making sure of the details. The full argument, together with the deduction of the complete intersection property, is given in [119].

In conclusion the key breakthrough in the proof had been the realization in the spring of 1991 that the two invariants introduced in the appendix could be used to relate the deformation rings and the Hecke rings. In effect the η-invariant could be used to count Galois representations. The last step after the June, 1993, announcement, though elusive, was but the conclusion of a long process whose purpose was to replace, in the ring-theoretic setting, the methods based on Iwasawa theory by methods based on the use of auxiliary primes.

One improvement that I have not included but which might be used to simplify some of Chapter 2 is the observation of Lenstra that the criterion for Gorenstein rings to be complete intersections can be extended to more general rings which are finite and free as $\mathbf{Z}_p$-modules. Faltings has pointed out an improvement, also not included,

Introduction 161

which simplifies the argument in Chapter 3 and [119]. This is however explained in the appendix to [119].

It is a pleasure to thank those who read carefully a first draft of some of this paper after the Cambridge conference and particularly N. Katz who patiently answered many questions in the course of my work on Euler systems, and together with Illusie read critically the Euler system argument. Their questions led to my discovery of the problem with it. Katz also listened critically to my first attempts to correct it in the fall of 1993. I am grateful also to Taylor for his assistance in analyzing in depth the Euler system argument. I am indebted to F. Diamond for his generous assistance in the preparation of the final version of this paper. In addition to his many valuable suggestions, several others also made helpful comments and suggestions especially Conrad, de Shalit, Faltings, Ribet, Rubin, Skinner and Taylor. Finally, I am most grateful to H. Darmon for his encouragement to reconsider my old argument. Although I paid no heed to his advice at the time, it surely left its mark.

Appendix B

List of Speakers Cambridge University June 1993

R. Taylor: Galois representations attached to automorphic forms over imaginary quadratic fields.
J-M. Fontaine: p-adic mixed Hodge structures.
B. Perrin-Riou: p-adic L-functions.
Y. Ihara: Galois lie algebras and higher commutators of Soulé elements.
P. Colmez: Values at $s = 0$ of p-adic L-functions.
U. de Shalit: p-adic regulators and K_2 of curves.
K. Ribet: Torsion points on $X_0(p)$.
R. Greenberg: Iwasawa theory for p-adic deformations.
U. Jannsen: On rigidity of K-cohomology.
K. Rubin: Some thoughts on Stark's conjecture.

Appendix B

M. Harrison: On the Block-Kato conjecture for Grossencharacters of $\mathbf{Q}(i)$.

P. Berthelot: On the Bernstein inequality for D^+-modules.

W. Messing: Remark on the f-ness of the image of the Abel-Jacobi map.

P. Schneider: Lattices in rational representations and sheaves on the Bruhat-Tits building.

J. Tilouine: Deformations of Galois representations.

S. Bloch: p-adic Arakelov theory.

B. Mazur: Global representations which are locally potentially crystalline.

A. Wiles: Modular forms, elliptic curves and Galois representations I, II, III.

Appendix C

List of Speakers Boston University August 1995

A. Invited Speakers

Wednesday August 9

9:00-10:00:	Glenn Stevens	Overview of the proof of Fermat's Last Theorem.
10:30-11:30:	Joe Silverman	Geometry of elliptic curves.
1:30-2:30:	Jaap Top* David Rohrlich	Modular curves.
3:00-4:00:	Larry Washington	Galois cohomology and Tate duality.

Appendix C

Thursday August 10

9:00-10:00:	Joe Silverman	Arithmetic of elliptic curves.
10:30-11:30:	Jaap Top* David Rohrlich	The Eichler-Shimura relations.
1:30-2:30:	John Tate	Finite group schemes.
3:00-4:00:	Jerry Tunnell* Steve Gelbart	Modularity of $\bar{\rho}_{E,3}$.

Friday August 11

9:00-10:00:	Dick Gross	Serre's Conjectures.
10:30-11:30:	Barry Mazur	Deformations of Galois representations: Introduction.
1:30-2:30:	Hendrik Lenstra	Explicit construction of deformation rings.
3:00-4:00:	Jerry Tunnell* Steve Gelbart	On the Langlands Program.

Saturday August 12

9:00-10:00:	Jerry Tunnell* Steve Gelbart	Proof of certain cases of Artin's Conjecture.
10:30-11:30:	Barry Mazur	Deformations of Galois representations: Examples.
1:30-2:30:	Dick Gross	Ribet's Theorem.
3:00-4:00:	Gerhart Frey	Fermat's Last Theorem and elliptic curves.

Monday August 14

9:00-10:00:	Jacques Tilouine	Hecke algebras and the Gorenstein property.

Appendix C 167

10:30-11:30:	René Schoof	The Wiles-Lenstra criterion for complete intersections.
1:30-2:30:	Barry Mazur	The tangent space and the module of Kähler differentials of the universal deformation ring.
3:00-4:00:	Ken Ribet	p-adic modular deformations of mod p modular representations.

Tuesday August 15

9:00-10:00:	René Schoof	The Wiles-Faltings criterion for complete intersections.
10:30-11:30:	Brian Conrad	The flat deformation functor.
1:30-2:30:	Larry Washington	Computations of Galois cohomology.
3:00-4:00:	Gary Cornell	Sociology, history and the first case of Fermat.

Wednesday August 16

9:00-10:00:	Ken Ribet	Wiles' "Main Conjecture."
10:30-11:30:	Ehud de Shalit	Modularity of the universal deformation ring (the minimal case).

Afternoon: free

Thursday August 17

9:00-10:00:	Alice Silverberg	Explicit families of elliptic curves with prescribed mod n representations.
10:30-11:30:	Ehud de Shalit	Estimating Selmer groups.
1:30-2:30:	Ken Ribet	Non-minimal deformations (the "induction step").
3:00-4:00:	Michael Rosen	Remarks on the history of Fermat's Last Theorem: 1844 to 1984.

Friday August 18

9:00-10:00:	Fred Diamond	An extension of Wiles' results.
10:30-11:30:	Karl Rubin	Modularity of mod 5 representations.
1:30-2:30:	Henri Darmon	Consequences and applications of Wiles' theorem on modular elliptic curves.
3:00-4:00:	Andrew Wiles	Modularity of semistable elliptic curves: Overview of the proof.

Note: * indicates the person who delivered the paper at the meeting.

B. Twenty Minute Talks, Sunday August 13.

9:30-9:50

A. William Ransdell — On Fermat's Last Theorem.
B. Gebhard Boeckle — Remarks on Deformations of Even Galois Representations.

10:00-10:20

A. Ali Yaakub — Introducing a Counterpart of Pascal Triangle.
B. Matthew Klassen — Algebraic Points on $X^5 + Y^5 = Z^5$ and an Extension of Fermat's Conjecture.
C. Boyd Roberts — Modularity of $\mathbf{Q}$ curves.

10:30-10:50

A. Malvina Baica — Baica's General Euclidean Algorithm (BGEA) and the Solution to Fermat's Last Theorem.
B. Li Guo — Selmer Groups and Special Values of Hecke L-Functions.
C. Tonghai Yang — Cusp Forms of Weight 1 Associated to Fermat Curves.

11:00-11:20

A. Edward Siegel — Attempt to Prove FLT via Dimensionality Calculus.
B. Pavlos Tzermias — Torsion in Mordell-Weil Groups of Fermat Jacobians.
C. Chad Schoen — Testing the Generalized Birch and Swinnerton-Dyer Conjecture.

Break

1:00-1:20

A. San Ling — On the Rational Cuspidal Subgroup and the Component Group.

B. Frasier Jarvis	Generalizations to Totally Real Fields.
C. Susan Goldstine	Canonical Heights and Polynomial Dynamics.

1:30-1:50

A. Nigel Boston	The Unramified Fontaine-Mazur Conjecture.
B. Joe Buhler	Searching For Irregular Pairs.
C. Yen-Mei Chen	The Selmer Groups and the Ambiguous Ideal Class Groups of Cubic Fields.

2:00-2:20

A. Lisa Fastenburg	The Rank of Some Infinite Families of Elliptic Surfaces.
B. Milja Poe	Elementary Proof of the First Case of FLT for Most Primes Congruent to -1 Modulo 3, that are Less than 500.
C. Bing Han	A Brief History (or Introduction to) of the Tamagawa Number Conjecture.

2:30-250

A. C. S. Rajan	On the Size of the Shafarevich-Tate Group of Elliptic Curves Over Function Fields.
B. Bart de Smit	A Differential Criterion for Complete Intersections.
C. Joan Lario	3-Torsion and Modularity.

3:00-3:30

A. Baxa and Schoissengeier	On the Distribution of the Sequence $(\propto \sqrt{n})$.
B. Vladimir Umanskiy	Some Properties of Prime Numbers.
C. Mohamed Sesay	The Connection Between Harmonic Analysis on Semi-Simple Lie Groups and Elliptic and Modular Curves.

Appendix D

Ram Murty's Review

This review was originally printed as a featured review in *Mathematical Reviews*: **99k:11004** *Modular forms and Fermat's Last Theorem.* Papers from the Instructional Conference on Number Theory and Arithmetic Geometry held at Boston University, Boston, MA, August 9–18, 1995. Edited by Gary Cornell, Joseph H. Silverman, and Glenn Stevens. Springer-Verlag, New York, 1997.

The story of Fermat's Last Theorem (FLT) and its resolution is now well-known. It is now common knowledge that Frey had the original idea linking the modularity of elliptic curves and FLT, that Serre refined this intuition by formulating precise conjectures, that Ribet proved a part of Serre's conjectures, which enabled him to establish that modularity of semistable elliptic curves implies FLT, and that finally Wiles proved the modularity of semistable elliptic curves.

The purpose of the book under review is to highlight and amplify these developments. As such, the book is indispensible to any student wanting to learn the finer details of the proof or any researcher wanting to extend the subject into a higher direction. Indeed, the subject is already expanding with the recent researches of Conrad, Darmon, Diamond, Skinner and others.

The book has 21 chapters. Chapters 18 and 19 are historical. Chapter 18 by Lenstra and Stevenhagen discusses class field theory and the first case of Fermat's Last Theorem. The early history (1844–1984) is given in Chapter 19 by Rosen, and it may be a good idea to begin the reading of this book with these chapters. The new ideas of the last two decades begin with Chapter 1 by Stevens, where an overview of the proof is given. The story begins with a possible solution of Fermat: $a^p + b^p = c^p$, and the construction of the corresponding Frey curve: $y^2 = x(x - a^p)(x + b^p)$, whose minimal discriminant is $\Delta = 2^{-8}(abc)^{2p}$ and conductor $N = \prod_{\ell | abc} \ell$. Since N is squarefree, this Frey curve is semistable.

What is the meaning of all these terms? Chapter 2 by Silverman covers the basic terminology and gives a brief survey of elliptic curves. Following a suggestion of Taniyama in 1957, Shimura and Weil, in the period between 1957 and 1967, formulated the moldularity conjecture more precisely. It says the following: Given any elliptic curve E over $\mathbf{Q}$, define $a_p(E)$ by setting it equal to $p + 1 - \#E(\mathbf{F}_p)$ where $\#E(\mathbf{F}_p)$ is the number of points on the curve mod p. (Technically, this is well-defined for all p coprime to N.) Then there exists a modular form $f(z)$ with Fourier expansion

$$f(z) = \sum_{n=1}^{\infty} a_n(f) e^{2\pi i n z}$$

such that $a_p(f) = a_p(E)$. Phrased another way, this says that the L-function of E coincides with the L-function of a modular form. This is to be viewed as a (nonabelian) reciprocity law. The relevant details about L-functions are found in Chapter 3 by Rohrlich.

Appendix D

Given a cusp form f of weight 2 and level N which is a normalized (i.e. $a_1(f) = 1$) eigenform of the Hecke operators, with $a_p(f)$ rational integers, Eichler and Shimura have described how one can construct an elliptic curve E_f such that we have the equality of the L-series: $L(E_f, s) = L(f, s)$. More precisely, fix $z_0 \in \mathbf{C}$ and consider the set

$$L_f = \left\{ \int_{z_0}^{\gamma z_0} f(z) dz : \quad \gamma \in \Gamma_0(N) \right\}.$$

It is not difficult to see that L_f is independent of the choice of $z_0 \in \mathbf{C}$. Shimura showed that L_f is a lattice in $\mathbf{C}$, the elliptic curve $E_f = \mathbf{C}/L_f$ is defined over $\mathbf{Q}$ and its L-function is equal to $L(f, s)$. The curves E_f are called modular elliptic curves and the modularity conjecture says that every elliptic curve over $\mathbf{Q}$ arises in this way.

The proof of FLT breaks up naturally into two parts: Ribet's theorem that the modularity conjecture for semistable elliptic curves implies FLT, and Wiles' theorem that every semistable elliptic curve over $\mathbf{Q}$ (and in particular, the above Frey curve) satisfies the modularity conjecture. To understand Ribet's theorem, one needs to understand Serre's conjectures, which are explained in Chapter 7 by Edixhoven. In a nutshell, Serre's conjectures predict that every continuous, irreducible, odd representation

$$\rho : \text{Gal}(\overline{\mathbf{Q}}/\mathbf{Q}) \to \text{GL}_2(\overline{\mathbf{F}}_\ell)$$

"comes from a cusp form f of weight $k(\rho)$ and level $N(\rho)$" whenever ℓ is a prime > 2. What this means is that

$$\text{tr}\rho(\text{Frob}_p) \equiv a_p(f) (\text{mod } \ell)$$

where Frob_p is the Frobenius automorphism. Serre makes a precise conjecture about what $k(\rho)$ and $N(\rho)$ should be. This is an amazing conjecture in itself and opens the way for a "Langlands program" mod p. Given an elliptic curve E over $\mathbf{Q}$, one can consider the action of $G_\mathbf{Q} = \text{Gal}(\overline{\mathbf{Q}}/\mathbf{Q})$ on the ℓ-division points $E[\ell]$ of E. Since $E[\ell]$ is a rank-2 $\mathbf{Z}/\ell\mathbf{Z}$ module, this gives us a representation

$$\rho_\ell : G_{\mathbf{Q}} \to \mathrm{GL}_2(\mathbf{F}_\ell).$$

By using Mazur's theorem on the classification of torsion of E over $\mathbf{Q}$, one can show that ρ_ℓ is irreducible for $\ell \geq 5$. By a theorem of Serre, this forces ρ_ℓ to be absolutely irreducible for $\ell \geq 5$. For $\ell = 3$, one has to modify this slightly, as Serre did when he formulated his conjectures, and this is important in the case of Wiles since $\ell = 3$ is the case he deals with initially.

Serre's prediction implies that the ρ_ℓ that arises from the action of $G_{\mathbf{Q}}$ on the ℓ-division points of the Frey curve constructed out of a possible solution of Fermat must come from a cusp form of level 2 and weight 2. Since there are no such forms, there is no nontrivial solution to FLT.

Serre's conjecture also implies the modularity conjecture. Indeed, given an elliptic curve E of conductor N, Serre's conjectures imply that for each prime ℓ there is an eigenform f_ℓ such that

$$a_p(E) = \mathrm{Tr}(\rho_\ell(\mathrm{Frob}_p)) \equiv a_p(f_\ell) (\mathrm{mod}\ \ell),$$

for all primes $\ell \geq 5$. Since there are only finitely many possibilities for f_ℓ, we must have, for some f, $f_\ell = f$ for infinitely many ℓ. Thus, for infinitely many ℓ, we have $\ell | (a_p(f) - a_p(E))$, which can only happen if $a_p(f) = a_p(E)$ for all primes p: this is the modularity conjecture.

Ribet proved a fundamental theorem about the level $N(\rho)$ in Serre's conjecture which validates the argument of the penultimate paragraph. This is explained in some detail in Chapter 7. The subject of finite flat group schemes arises in both Ribet's work and Wiles's work, so this is expounded in Chapter 5 by Tate.

According to theorems which are due to Serre and Faltings, two elliptic curves are isogenous if and only if their ℓ-adic representations arising from the action of $G_{\mathbf{Q}}$ on the Tate module $V_\ell(E) = \varprojlim E[\ell^n] \otimes \mathbf{Q}_\ell$ are isomorphic for some prime ℓ. This is the same as saying that the corresponding L-functions are equal. Thus, to show that E is

Appendix D 175

modular, it "suffices" to find a modular elliptic curve E' such that the corresponding ℓ-adic representations are isomorphic for some prime ℓ. Wiles chooses $\ell = 3$ and begins by considering the action of $G_\mathbf{Q}$ on $E[3]$. This gives a representation of $G_\mathbf{Q}$ into $\text{PGL}_2(\mathbf{F}_3)$ which is isomorphic to S_4. By a celebrated theorem of Langlands and Tunnell, there is a modular form g of weight one such that $\text{Tr}\rho_3(\text{Frob}_p) = a_p(g)$, which means that $a_p(E) \equiv a_p(g) (\text{mod } 3)$. Since the weight is "wrong" Wiles multiplies g by an Eisenstein series E_1 of weight one (with Nebentypus) such that $E_1 \equiv 1 (\text{mod } 3)$. This gives a form f of weight 2 which can be lifted to an eigenform by a theorem of Deligne. This at least proves the modularity conjecture mod 3! For this, one needs to use one of the central results in the Langlands program. The object of Chapter 6 by Gelbart is to describe the proof of the important Langlands-Tunnell theorem.

If one can extend the congruence modulo higher powers of 3, then the modularity conjecture follows. This is where Mazur's universal deformation ring enters the picture. Let G_S be the Galois group of the maximal extension of $\mathbf{Q}$ unramified outside S. Let $\overline{\rho} : G_S \to \text{GL}_2(\mathbf{F}_\ell)$ be given. Suppose $\overline{\rho}$ is absolutely irreducible. Suppose further that there is a complete local Noetherian ring A whose residue field is $\mathbf{F}_\ell$ and such that ρ lifts to $\rho_A : G_S \to \text{GL}_2(A)$ and $\rho = \rho_A \circ \text{pr}$ where pr is the natural projection of A onto $\mathbf{F}_\ell$. Among all such A, Mazur showed that there is a "universal one" in the following sense. There is a complete Noetherian local ring R with residue field $\mathbf{F}_\ell$ which lifts $\overline{\rho}$ to $\rho_R : G_S \to \text{GL}_2(R)$, which commutes with the natural projection of $\text{GL}_2(R) \to \text{GL}_2(\mathbf{F}_\ell)$ and which satisfies the following universal property: For every representation $\rho_A : G_S \to \text{GL}_2(A)$ there is a unique homomorphism $\pi : R \to A$ such that $\rho_A = \rho_R \cdot \pi$. R is called the universal deformation ring. In Wiles' proof, one needs to impose additional conditions on R. These technical details are discussed in Chapter 8 by Mazur and Chapter 9 by de Smit and Lenstra.

Some of these lifts ρ_R arise from modular forms (as is suggested by the Eichler-Shimura construction, for example). These modular lifts of the given $\bar{\rho}$ give a representation into $\mathrm{GL}_2(T)$, where T is the Hecke ring. By the universal property, T is a quotient of R. The main theorem of Wiles shows that $R = T$, that is, that all lifts are modular. To this end, he needs to use many deep properties about the Hecke ring T, the main one being that T is Gorenstein. One of the features of the Wiles' proof is his simple numerical criterion for $R = T$ in terms of Galois cohomology. The preliminary discussion on Galois cohomology is done in Chapter 4 by Washington. The discussion of the Gorenstein property is done in Chapter 10 by Tilouine, Chapter 11 by de Smit, Rubin and Schoof, Chapter 12 by Diamond and Ribet, Chapter 13 by Conrad and Chapter 14 by de Shalit. These chapters form the major portion of the text.

All of this discussion applies if ρ_3 is irreducible. If it is reducible, Wiles considers the action of $G_\mathbf{Q}$ on the 5-division points. By a clever use of the Chebotarev density theorem and Hilbert's irreducibility theorem Wiles shows that there is a semistable curve E' such that $\rho_{5,E} \simeq \rho_{5,E'}$ and $\rho_{3,E'}$ is irreducible. Thus E' is modular. But then, $\rho_{5,E} \simeq \rho_{5,E'}$ implies that E and E' are isogenous by the Serre-Faltings theorem. These ideas, as well as the modularity of $\mathbf{F}_4$ and $\mathbf{F}_5$ representations, are discussed in Chapter 15 by Silverberg and Chapter 16 by Rubin.

In Chapter 17, Diamond extends the work of Wiles to prove the modularity of elliptic curves which are semistable at 3 and 5. Very recently, Conrad, Diamond, Taylor and Christophe Breuil have proved that *all* elliptic curves over the rationals are modular. Their methods are further developments of the approach outlined in the articles of this volume. In Chapter 20, Frey discusses the generalised Fermat equation. Finally, in the last chapter, Darmon discusses the connection of Wiles' theorem to the Birch and Swinnerton-Dyer conjectures about the order of zeros at the central critical point.

The book has a nicely prepared index and with its numerous references is a welcome addition to the library of every student and researcher in the area. The student may also consult the related

Appendix D

volume [*Seminar on Fermat's Last Theorem*, (Toronto, ON, 1993–1994), Amer. Math. Soc., Providence, RI, 1995; MR 96f:11004], which appeared several years earlier and attempted to achieve the same goal of giving a detailed overview of the Wiles' proof.

FLT deserves a special place in the history of civilization. Because of its simplicity, it has tantalized amateurs and professionals alike, and its remarkable fecundity has led to the development of large areas of mathematics such as, in the last century, algebraic number theory, ring theory, algebraic geometry, and in this century, the theory of elliptic curves, representation theory, Iwasawa theory, formal groups, finite flat group schemes and deformation theory of Galois representations, to mention a few. It is as if some supermind planned it all and over the centuries had been developing diverse streams of thought only to have them fuse in a spectacular synthesis to resolve FLT. No single brain can claim expertise in all of the ideas that have gone into this "marvelous proof". In this age of specialization, where "each one of us knows more and more about less and less", it is vital for us to have an overview of the masterpiece such as the one provided by this book.

Bibliography

1. A. Aczel, *Fermat's Last Theorem*, Four Walls Eight Windows, New York, 1996.
2. A. Altman and S. Kleiman, *An Introduction to Grothendieck Duality Theory*, vol. 146, Springer Lecture Notes in Mathematics, 1970.
3. B. Birch and W. Kuyk (eds.), *Modular Functions of One Variable IV*, vol. 476, Springer Lecture Notes in Mathematics, 1975.
4. S. Bloch and K. Kato, *L-Functions and Tamagawa Numbers of Motives*, in The Grothendieck Festschrift, vol. 1 (P. Cartier et al., eds.), Birkhäuser, 1990.
5. N. Boston, *Families of Galois representations—Increasing the ramification*, Duke Math. J. **66** (1992), 357–367.
6. N. Boston, H. Lenstra, and K. Ribet, *Quotients of group rings arising from two-dimensional representations*, C. R. Acad. Sci. Paris Sér. I **312** (1991), 323–328.
7. W. Bruns and L. Herzog, *Cohen-Macaulay Rings*, Cambridge University Press, 1993.
8. H. Carayol, *Sur les représentations p-adiques associées aux formes modulaires de Hilbert*, Ann. Sci. Ec. Norm. Sup. (4) **19** (1986), 409–468.
9. H. Carayol, *Sur les représentations galoisiennes modulo attachées aux formes modulaires de Hilbert*, Duke Math. J. **59** (1989), 785–801.

10. H. Carayol, *Formes modularies et représentations Galoisiennes à valeurs dans un anneau local complet*, in p-Adic Monodromy and the Birch-Swinnerton-Dyer Conjecture (B. Mazur and G. Stevens, eds.), Contemp. Math., vol. 165, 1994.
11. J. W. S. Cassels and A. Frölich (eds.), *Algebraic Number Theory*, Academic Press, 1967.
12. E. Cline, B. Parshall, and L. Scott, *Cohomology of finite groups of Lie type I*, Inst. Hautes Études Sci. Publ. Math. **45** (1975), 169–191.
13. J. Coates and C. G. Schmidt, *Iwasawa theory for the symmetric square of an elliptic curve*, J. Reine und Angew. Math. **375/376** (1987), 104–156.
14. J. Coates and A. Wiles, *On p-adic L-functions and elliptic units*, J. Aust. Math. Soc. Ser. A **26** (1978), 1–25.
15. J. Coates and S. T. Yau, *Elliptic Curves, Modular Forms & Fermat's Last Theorem*, 2nd ed., International Press Incorporated, Cambridge, MA, 1997.
16. R. Coleman, *Division values in local fields*, Invent. Math. **53** (1979), 91–116.
17. B. Conrad, F. Diamond, and R. Taylor, *Modularity of certain potentially Barsotti-Tate Galois representations*, Jour. Amer. Math. Soc. **12** (1999), 521–567.
18. G. Cornell and J. H. Silverman (eds.), *Arithmetic Geometry*, Springer-Verlag New York, Inc., 1986.
19. G. Cornell, J. H. Silverman, and G. Stevens, (eds.), *Modular Forms and Fermat's Last Theorem*, Springer-Verlag New York, Inc., 1998.
20. D. Cox, *Introduction to Fermat's Last Theorem*, Amer. Math. Monthly **101** (1994), 3–14.
21. P. Deligne, *La conjecture de Weil, I*, Inst. Hautes Études Sci. Publ. Math. **43** (1974), 273–307.
22. P. Deligne, *La conjecture de Weil, II*, Inst. Hautes Études Sci. Publ. Math. **52** (1980), 137–252.
23. P. Deligne and M. Rapoport, *Schémes de modules de courbes elliptiques*, vol. 349, Springer Lecture Notes in Mathematics, 1973.
24. P. Deligne and J-P. Serre, *Formes modulaires de poids 1*, Ann. Sci. Ec. Norm. Sup. (4) **7** (1974), 507–530.

25. F. Diamond, *The refined conjecture of Serre*, in Elliptic Curves, Modular Forms and Fermat's Last Theorem, 2nd ed. (J. Coates and S. T. Yau, eds.), International Press Incorporated, Cambridge, MA, 1997, pp. 172–186.

26. F. Diamond and K. A. Ribet, *ℓ-adic modular deformations and Wiles's "Main Conjecture"*, in Modular Forms and Fermat's Last Theorem, Springer-Verlag New York, Inc., 1998, pp. 357–371.

27. F. Diamond and R. Taylor, *Lifting modular mod l representations*, Duke Math. J. **74** (1994), 253–269.

28. L. E. Dickson, *Linear Groups with an Exposition of the Galois Field Theory*, Teubner, Leipzig, 1901.

29. J. Dieudonne, *A Panorama of Pure Mathematics*, Academic Press, Inc., New York, 1982.

30. V. Drinfeld, *Two-dimensional ℓ-adic representations of the fundamental group of a curve over a finite field and automorphic forms on* GL(2), Am. J. Math. **105** (1983), 85–114.

31. B. Edixhoven, *L'action de l'algèbre de Hecke sur les groupes de composantes des jacobiennes des courbes modulaires est "Eisenstein"*, in Courbes Modulaires et Courbes de Shimura, **196–197** (1991), pp. 159–170.

32. B. Edixhoven, *The weight in Serre's conjecture on modular forms*, Invent. Math. **109** (1992), 563–594.

33. H. M. Edwards, *Riemann's Zeta Function*, Academic Press, New York, 1974.

34. H. M. Edwards, *Fermat's Last Theorem*, Springer-Verlag, New York, 1977.

35. M. Flach, *A finiteness theorem for the symmetric square of an elliptic curve*, Invent. Math. **109** (1992), 307–327.

36. J.-M. Fontaine, *Sur certains types de représentations p-adiques du groupe de Galois d'un corps local; construction d'un anneau de Barsotti-Tate*, Ann. of Math. **115** (1982), 529–577.

37. J. M. Fontaine and G. Lafaille, *Construction de représentations p-adiques*, Ann. Sci. Ec. Norm. Sup. **15** (1982), 547–608.

38. G. Frey, *Links between stable elliptic curves and certain diophantine equations*, Annales Universitatis Saraviensis **1** (1986), 1–40.

39. S. Gelbart, *An elementary introduction to the Langlands program*, Bull. Amer. Math. Soc. **10** (1984), 177–219.
40. D. Goldfeld, *Beyond the Last Theorem*, The Sciences, March/April (1996), 34–40.
41. F. Gouvêa, *A marvelous proof*, Amer. Math. Monthly **101** (1994), 203–222.
42. R. Greenberg, *On the structure of certain Galois groups*, Invent. Math. **47** (1978), 85–99.
43. R. Greenberg, *Iwasawa theory for p-adic representations*, Adv. St. Pure Math. **17** (1989), 97–137.
44. B. Gross, *A tameness criterion for Galois representations associated to modular forms* mod p, Duke Math. J. **61** (1990), 445–517.
45. L. Guo, *General Selmer groups and critical values of Hecke L-functions*, Math. Ann. **297** (1993), 221–233.
46. J. Hadamard, *The Psychology of Invention in the Mathematical Field*, Princeton University Press, 1945,1949; reprinted (1949 enlarged edition) Dover Publications, Inc., New York, 1954.
47. Y. Hellegouarch, *Points d'ordre $2p^h$ sur les courbes elliptiques*, Acta Arith. **26** (1975), 253–263.
48. H. Hida, *Congruences of cusp forms and special values of their zeta functions*, Invent. Math. **63** (1981), 225–261.
49. H. Hida, *Iwasawa modules attached to congruences of cusp forms*, Ann. Sci. Ecole Norm. Sup. (4) **19** (1986), 231–273.
50. H. Hida, *On p-adic Hecke algebras for* GL_2 *over totally real fields*, Ann. of Math. **128** (1988), 295–384.
51. H. Hida, *Theory of p-adic Hecke algebras and Galois representations*, Sugaku Expositions **2-3** (1989), 75–102.
52. B. Huppert, *Endliche Gruppen I*, Springer-Verlag, 1967.
53. Y. Iharra, *On modular curves over finite fields*, in Proc. Intern. Coll. on discrete subgroups of Lie groups and application to moduli, Bombay, 1973, pp. 161–202.
54. H. Iwaniec and P. Sarnak, *The non-vanishing of central values of automorphic L-functions and Siegel's zeros*, Israel Journal of Mathematics (to appear).

Bibliography

55. H. Iwaniec and P. Sarnak, *Analytic theory of L-functions*, Proceedings of the Conference "Visions of Mathematics Toward 2000", Tel Aviv University (to appear).
56. K. Iwasawa, *On $\mathbf{Z}_l$-extensions of algebraic number fields*, Ann. of Math. **98** (1973), 246–326.
57. H. Jacquet, I. I. Piatetski-Shapiro, and J. Shalika, *Relèvement cubique non normal*, C. R. Acad. Sci. Paris **292** (1981), 567–571.
58. N. Katz, *A result on modular forms in characteristic p*, in Modular Functions of One Variable V, vol. 601, Springer Lecture Notes in Mathematics (1976), 53–61.
59. N. Katz and B. Mazur, *Arithmetic Moduli of Elliptic Curves*, Ann. of Math. Studies, **108**, Princeton University Press, 1985.
60. N. Katz and P. Sarnak, *Zeroes of zeta functions and symmetry*, Bull. Amer. Math. Soc. **36** (1999), 1–26.
61. N. Katz and P. Sarnak, *Random Matrices, Frobenius Eigenvalues and Monodromy*, American Mathematical Society, Providence, RI, 1999.
62. E. Kunz, *Almost complete intersections are not Gorenstein*, J. Alg. **28** (1974), 111–115.
63. E. Kunz, *Introduction to Commutative Algebra and Algebraic Geometry*, Birkhäuser, 1985.
64. S. Lang, *Introduction to Arakelov Theory*, Springer-Verlag New York, Inc., 1988.
65. S. Lang, *Cyclotomic Fields I and II*, 2nd edition, Springer-Verlag New York, Inc., 1990.
66. R. Langlands, *Base Change for* GL(2), Ann. of Math. Studies **96**, Princeton University Press, 1980.
67. H. Lenstra, *Complete intersections and Gorenstein rings*, in Elliptic Curves, Modular Forms and Fermat's Last Theorem, 2nd ed. (J. Coates and S. T. Yau, eds.), International Press Incorporated, Cambridge, MA, 1997, pp. 248–257.
68. W. Li, *Newforms and functional equations*, Math. Ann. **212** (1975), 285–315.
69. R. Livné, *On the conductors of mod ℓ Galois representations coming from modular forms*, J. of No. Th. **31** (1989), 133–141.
70. B. Mazur, *Modular curves and the Eisenstein ideal,* Inst. Hautes Études Sci. Publ. Math. **47** (1977), 133–186.

71. B. Mazur, *Rational isogenies of prime degree*, Invent. Math. **44** (1978), 129–162.
72. B. Mazur, *Deforming Galois representations*, in Galois Groups over **Q**, vol. 16, MSRI Publications, Springer, New York, 1989.
73. B. Mazur, *Fermat's Last Theorem*, Selected Lectures in Mathematics, AMS, Providence, RI, 1995, video recording.
74. B. Mazur and K. Ribet, *Two-dimensional representations in the arithmetic of modular curves, Courbes Modulaires et Courbes de Shimura*, Astérisque **196-197** (1991), 215–255.
75. B. Mazur and L. Roberts, *Local Euler characteristics*, Invent. Math. **9** (1970), 201–234.
76. B. Mazur and J. Tilouine, *Représentations galoisiennes, différentielles de Kähler et conjectures principales,* Inst. Hautes Études Sci. Publ. Math. **71** (1990), 65–103.
77. B. Mazur and A. Wiles, *Class fields of abelian extensions of* **Q**, Invent. Math. **76** (1984), 179–330.
78. B. Mazur and A. Wiles, *On p-adic analytic families of Galois representations*, Comp. Math. **59** (1986), 231–264.
79. J. S. Milne, *Jacobian varieties*, in Arithmetic Geometry (Cornell and Silverman, eds.), Springer-Verlag New York, Inc., 1986.
80. J. S. Milne, *Arithmetic Duality Theorems*, Academic Press, 1986.
81. V. K. Murty (ed.), *Seminar on Fermat's Last Theorem*, CMS Conference Proceedings, American Mathematical Society, **17**, Providence, RI, 1995.
82. T. Nagell, *Introduction to Number Theory*, Wiley, New York, (1951); reprinted by Chelsea, New York, 1962.
83. A. van der Poorten, *Notes on Fermat's Last Theorem*, John Wiley & Sons, Inc., New York, 1996.
84. R. Ramakrishna, *On a variation of Mazur's deformation functor*, Comp. Math. **87** (1993), 269–286.
85. M. Raynaud, *Spécialisation du foncteur de Picard,* Inst. Hautes Études Sci., Publ. Math. **38** (1970), 27–76.
86. M. Raynaud, *Schémas en groupes de type* $(p, p, \ldots, p)$, Bull. Soc. Math. France **102** (1974), 241–280.

Bibliography

87. P. Ribenboim, *13 Lectures on Fermat's Last Theorem*, Springer-Verlag New York, Inc., 1979.
88. P. Ribenboim, *Fermat's Last Theorem for Amateurs*, Springer-Verlag New York, Inc., 1999.
89. K. A. Ribet, *Congruence relations between modular forms,* Proc. Int. Cong. of Math. **17** (1983), 503–514.
90. K. A. Ribet, *From the Taniyama-Shimura conjecture to Fermat's Last Theorem*, Ann. Fac. Sci. Toulouse Math. **11** (1990), 116–139.
91. K. A. Ribet, *On modular representations of* $\mathrm{Gal}(\bar{\mathbf{Q}}/\mathbf{Q})$ *arising from modular forms*, Invent. Math. **100** (1990), 431–476.
92. K. A. Ribet, *Multiplicities of p-finite mod p Galois representations in* $J_0(N_p)$, Boletin da Sociedade Brasileira de Matematica, Nova Serie **21** (1991), 177–188.
93. K. A. Ribet, *Wiles proves Taniyama's conjecture; Fermat's Last Theorem follows*, Notices of Amer. Math. Soc. **40** (1993), 575–576.
94. K. A. Ribet, *Modular Elliptic Curves and Fermat's Last Theorem*, Selected Lectures in Mathematics, AMS, Providence, RI, 1993, video recording.
95. K. A. Ribet, *Report on* mod l *representations of* $\mathrm{Gal}(\bar{\mathbf{Q}}/\mathbf{Q})$, Proc. of Symp. in Pure Math., vol. 55, 1994, pp. 639–676.
96. K. A. Ribet, *Galois representations and modular forms*, Bull. Amer. Math. Soc. **32** (1995), 375–402.
97. K. A. Ribet, *Erratum to "Galois representations and modular forms"*, Bull. Amer. Math Soc. **33** (1996), 43.
98. K. A. Ribet and B. Hayes, *Fermat's Last Theorem and modern arithmetic*, American Scientist (March-April, 1994), 144–156.
99. K. Rubin, *Elliptic curves with complex multiplication and the conjecture of Birch and Swinnerton-Dyer*, Invent. Math. **64** (1981), 455–470.
100. K Rubin, *Tate-Shafarevich groups and L-functions of elliptic curves with complex multiplication*, Invent. Math. **89** (1987), 527–560.
101. K. Rubin, *The 'main conjectures' of Iwasawa theory for imaginary quadratic fields*, Invent. Math. **103** (1991), 25–68.
102. K. Rubin, *More 'main conjectures' for imaginary quadratic fields*, in CRM Proceedings and Lecture Notes **4** (1994).

103. K. Rubin and A. Silverberg, *A report on Wiles' Cambridge lectures*, Bull. Amer. Math. Soc. (N.S.) **31** (1994), 15–38; Erratum **32** (1995), 170.

104. P. Sarnak et al. (eds.), *A Celebration of John F. Nash, Jr.*, Duke University Press, 1996.

105. M. Schlessinger, *Functors on Artin rings*, Trans. A.M.S. **130** (1968), 208–222.

106. R. Schoof, *The structure of the minus class groups of abelian number fields Seminaire de Théorie des Nombres, Paris (1988–1989)*, Progress in Math. **91** (1990), Birkhäuser, 185–204.

107. J-P. Serre, *Sur les représentations modulaires de degré* 2 *de* $\mathrm{Gal}(\bar{\mathbf{Q}}/\mathbf{Q})$, Duke Math. J. **54** (1987), 179–230.

108. J.-P. Serre, *Lectures on the Mordell-Weil Theorem*, Friedr. Vieweg & Son, Braunschweig/Wiesbaden, 1989.

109. E. de Shalit, *Iwasawa Theory of Elliptic Curves with Complex Multiplication*, Persp. Math., vol. 3, Academic Press, 1987.

110. E. de Shalit, *On certain Galois representations related to the modular curve* $X_1(p)$, Comp. Math. **95** (1995), 69–100.

111. G. Shimura, *Introduction to the Arithmetic Theory of Automorphic Functions*, Iwanami Shoten and Princeton University Press, 1971.

112. G. Shimura, *On elliptic curves with complex multiplication as factors of the Jacobians of modular function fields*, Nagoya Math. J. **43** (1971), 199–208.

113. G. Shimura, *On the holomorphy of certain Dirichlet series*, Proc. London Math. Soc., (3) **31** (1975), 79–98.

114. G. Shimura, *The special values of the zeta function associated with cusp forms*, Comm. Pure and Appl. Math., **29** (1976), 783–804.

115. G. Shimura, *Yutaka Taniyama and his time: very personal recollections*, Bull. Lond. Math. Soc. **21** (1989), 186–196.

116. S. Singh, *Fermat's Enigma*, Walker and Company, New York, 1997.

117. S. Singh and K.A. Ribet, *Fermat's last stand*, Scientific American (November, 1997), 68–73.

118. C. Soulé et al., *Lectures on Arakelov Geometry*, Cambridge University Press, Cambridge, 1982.

Bibliography

119. R. Taylor and A. Wiles, *Ring-theoretic properties of certain Hecke algebras*, Ann. of Math. **141** (1995), 553–572.
120. J. Tilouine, *Un sous-groupe p-divisible de la jacobienne de $X_1(Np^r)$ comme module sur l'algébre de Hecke*, Bull. Math. Soc. France **115** (1987), 329–360.
121. J. Tilouine, *Théorie d'Iwasawa classique et de l'algébre de Hecke ordinaire*, Comp. Math. **65** (1988), 265–320.
122. J. Tunnell, *Artin's conjecture for representations of octahedral type*, Bull. Amer. Math. Soc. **5** (1981), 173–175.
123. J. Tunnell, *A classical diophantine problem and modular forms of weight 3/2*, Invent. Math. **72** (1983), 323–334.
124. L. C. Washington, *Introduction to Cyclotomic Fields*, 2nd edition, Springer-Verlag New York, Inc., 1997.
125. A. Weil, *Über die Bestimmung Dirichletcher Reihen durch Funktionalgleichungen*, Math. Ann. **168** (1967), 149–156.
126. A. Wiles, *Modular curves and the class group of $Q(\zeta_p)$*, Invent. Math. **58** (1980), 1–35.
127. A. Wiles, *On p-adic representations for totally real fields*, Ann. of Math. **123** (1986), 407–456.
128. A. Wiles, *On ordinary λ-adic representations associated to modular forms*, Invent. Math. **94** (1988), 529–573.
129. A. Wiles, *The Iwasawa conjecture for totally real fields*, Ann. of Math. **131** (1990), 493–540.
130. A. Wiles, *Modular elliptic curves and Fermat's Last Theorem*, Ann. of Math. **142** (1995), 443–551.
131. J. P. Wintenberger, *Structure galoisienne de limites projectives d'unitées locales*, Comp. Math. **42** (1981), 89–103.

Index

ABC conjecture, 11
Aczel, Amir, 6, 141
arithmetic surfaces, 11
Arithmetica, ix, 147
Arthur, James, 53, 60, 69
Barner, Klaus, 99, 137
Bell, Eric Temple, 4
Birch, Bryan, 71
Blum, Lenore, 3
Bombieri, Enrico, 2, 3, 65, 140
Borel, Armand, 139, 144
Boston University, x, 97
Boston University Conference, x, 97
Breuil, Christophe, 101, 102, 121, 137
Bumby, Justine, 57
Cambridge, England, 1, 18, 80
Cambridge University, 17, 18, 61, 71, 80

Cambridge University Conference, 2, 16, 17, 20, 21, 54
Cartier-Bresson, Henri, 77
Chinese University of Hong Kong, 4
Chowla, Sarva Daman, 77, 144
Claflin Hall, 99
class number formula, 15
Coates, John, 4, 15, 17, 18, 33, 72
Columbia University, x, 76
Conrad, Brian, 23, 24, 45, 60, 70, 100, 101
Conway, John, 3, 24
Cornell, Gary, 98, 119
Cox, David, 4, 140
Darmon, Henri, 23, 24, 39, 141
deformations of Galois representations, 11, 14
Deligne, Pierre, 23, 38, 147
Diamond, Fred, 18, 23, 24, 40, 67, 68, 101, 102
Diophantus of Alexandria, ix, 147
Eisenstein ideal, 8
elliptic curves, x, 4, 6–11, 13, 17, 25, 54, 68, 78, 97, 139
epsilon conjecture, 13
euler products, 14
euler system, 54, 55
Faltings, Gerd, 22, 23, 26, 36, 37, 58, 64, 68, 72, 73, 77
Fermat, Pierre de, ix, 60, 143, 147
Fermat's Last Theorem, ix, x, 1, 2, 4, 6–10, 14–16, 19–21, 26, 54, 56, 57, 59, 65, 67, 74, 77, 78, 97, 98, 140, 141, 143, 148, 149
Fields Institute for Research in Mathematical Sciences, 4
Fields, John Charles, 145
Fields Medal, 25, 72, 140, 145, 147, 149

Index

Fine Hall, ii, 2, 21, 23, 25, 58
Flach, Matthias, 14, 15, 55, 56, 64, 66
Fontaine, Jean-Marc, 31, 149
Freedom of Information Act, 15
Frey curves, 9
Frey, Gerhard, x, 5, 6, 7, 11, 97, 98, 100, 101, 104
Galois module structures, 8
Galois representations, 12, 13, 98
generalized Riemann hypothesis, 147
Goldfeld, Dorian, 76, 122, 137, 138, 142, 143, 144
Gouvêa, Fernando, 4, 140
Greenberg, Ralph, 14, 17, 18
Griffiths, Phillip, 38
Gross, Benedict, 113, 142
Grothendieck, John, 57
Guo, Li, 17, 137, 143
Hadamard, Jacques, 57
Hadamard's rule, 57
Hardy, G. H., 57
Harvard University, x, 61, 72, 80
Hearst, Will, 3
Hecke algebras, 54, 98
Hecke rings, 67
Hellegouarch, Yves, 6, 7, 8, 103
Hermann Weyl Lectures, 53
Hunter College, x, 61
Illusie, Luc, 19
Institute for Advanced Study, 1, 25, 53, 60, 61, 72, 80, 102, 139

Instructional Conference on Number Theory and Arithmetic Geometry, x, 97, 171
International Congress of Mathematicians, 144, 149
Inventiones Mathematicae, 12, 19
Isaac Newton Institute for Mathematical Sciences, 17
Iwaniec, Henryk, 137
Iwasawa theory, 15, 56, 57
j-invariant, 20, 25, 144
Jackson, Allyn, 25, 141
Jacob Sleeper Hall, 97
Jacquet, Hervé, 23, 47, 137
John D. and Catherine T. MacArthur Foundation, 148
Joint Policy Board of Mathematics (JPBM), 142
Katz, Ian, 70
Katz, Nicholas, 14, 19, 21, 22, 23, 29, 64
King Faisal Foundation, 149
King Faisal International Prize for Science, 80, 149
Kohn, Joseph J., 58, 60, 76, 137
Kolata, Gina, 19, 26, 62, 142
Kolyvagin–Flach Method, 14, 55, 56
Kolyvagin, Victor, 15, 55, 56
Kummer, Ernst Eduard, 9, 148
Kummer's Theory, 9
Lang, Serge, 138
Langlands, Robert, 23, 50, 60, 61, 69, 80, 144, 148
Langlands philosophy, 9
Langlands program, 69, 70, 144
Langlands-Tunnell result, 61
Lenstra, Hendrik, 111

Index

level-lowering principle, 5, 12
Lindemann, Ina, 100
Lynch, John, 142
MacArthur Fellowship, 80, 148
MacPherson, Robert, 61
Main Conjecture of Iwasawa, 72
Mathematical Research Institute Oberwolfach, ix, 5, 6, 9, 10, 141
Mathematical Sciences Research Institute (MSRI), 2, 3, 140
Mazur, Barry, 3, 8, 11, 14, 16, 18, 19, 30, 61, 74, 106
Mazur's Theorem, 7, 8
Mestre, Jean-François, 12
method of infinite descent, 74
Mittag-Leffler, Gosta, 145
Mittag-Leffler Institute, 145
modular forms, 4, 12, 13, 78, 98
Mordell's conjecture, 59
Murty, Kumar, 4
Murty, Ram, 98, 123, 137, 171
Nash, John, 60, 145
National Academy of Science Award in Mathematics, 80, 148
National Science Foundation, 14
New York Times, 1, 21, 26, 53, 62, 63
New York University, x, 75
Nobel, Alfred, 145
Nobel Prize, 145
NOVA broadcast, 142
Ogg's Conjecture, 7
Osserman, Robert, 3
Oxford University, 19, 71, 80

Piatetski-Shapiro, Ilya, 23, 48
pigeonhole principle, 67
Princeton, x, 1, 20, 72
Princeton home, 14, 16, 54, 96
Princeton University, 72, 80
Prix Fermat, 12, 80, 143
Ribet, Kenneth, x, 2, 3, 5, 9, 11, 12, 13, 18, 19, 64, 77, 78, 97, 98, 100, 101, 105, 137, 138, 140, 142
Riemann hypothesis, 62, 65, 147
Roitberg, Joseph, 61, 146
Rosen, Michael, 118
Rubin, Karl, 2, 3, 4, 54, 55, 112, 137, 140
Rutgers University, x, 77, 79
Sarnak, Peter, 16, 20, 23, 25, 32, 137
Schock Prize, 80, 147
Schoof, René, 108
Scientific American, 21, 26
Selmer group, 14, 19, 22, 78, 98
semistable elliptic curves, x, 9, 13, 25, 68
Serre's Conjectures, 9, 10, 11, 12
Serre, Jean-Pierre, x, 5, 8, 9, 12, 34, 35, 97, 98, 100, 142
Shalika, Joseph A., 23, 49
de Shalit, Ehud, 18, 109
Shapiro, Harold, 2
Shimura, Goro, 81, 138
Shimura-Taniyama Conjecture, x, 5, 9, 10, 13, 14, 17, 24, 61, 69, 78, 97, 101, 138, 139, 149
Siegel, Carl Ludwig, 99, 139
Silverberg, Alice, 4, 117, 140
Silverman, Joseph, 98, 114

Index

Singh, Simon, 6, 142
Skinner, Chris, 23, 24, 44, 70, 100
Springer-Verlag, 98, 100
Steele, Leroy P., 146
Steele Prize, 23, 146
Stevens, Glenn, 98, 115
Swinnerton-Dyer, Peter, 71, 72
symmetric square representation of a modular form, 14, 19, 22
Szpiro, Lucien, 11, 43
Taniyama, Yutako, 139
Taplin Auditorium, 21, 22, 24, 57, 60, 102
Tate, John, 35, 107
Taylor, Richard, x, 24, 54, 55, 57, 58, 60, 62, 63, 65, 66, 67, 68, 86, 98, 99, 100, 101, 141
The Daily Princetonian, 60
The Guardian, 70
The Last Problem, 4
Tilouine, Jacques, 110
Top, Jaap, 116
torsion points, 7, 8
Tunnell, Jerrold, 23, 24, 46, 61, 77, 78, 79, 144
Vatsal, Nike, 23, 44
Washington, Lawrence, 23, 24, 41, 99, 137, 138
Weil, André, 20, 26, 81, 138, 139
Wiles, Alix, 71, 74
Wiles, Andrew, x, 1–4, 13–28, 37, 51, 53–55, 57–80, 82–85, 87–95, 97, 98, 100, 101, 102, 120, 138, 140, 143, 146, 148, 149
Wiles, Maurice, 71
Wiles, Nada, 16, 56, 58, 76
Wiles-Taylor collaboration, 24, 63

Wiles's attic office, 14, 16, 53, 96
Wolf Prize, 80, 148
Wolfskehl, Paul, 148
Wolfskehl Prize, 80, 148
Yale University, x, 75
Yau, Shing-Tung, 4
Zhang, Shou-Wu, 14, 23, 24, 42, 137, 143

QA 244 .M69 2000
Mozzochi, Charles J.
The Fermat diary

DATE DUE

AUG 1 6 2002			
DEC 2 0 2003			